Nature Technology

Emile H. Ishida • Ryuzo Furukawa

Nature Technology

Creating a Fresh Approach to Technology and Lifestyle

 Springer

Emile H. Ishida
Tohoku University
Sendai, Miyagi, Japan

Ryuzo Furukawa
Tohoku University
Sendai, Miyagi, Japan

ISBN 978-4-431-56161-3   ISBN 978-4-431-54613-9 (eBook)
DOI 10.1007/978-4-431-54613-9
Springer Tokyo Heidelberg New York Dordrecht London

Printed on acid-free paper

Springer is part of Springer Science+Business Media (www.springer.com)

Preface

Starting in 1978, I worked 26 years for the INAX Corporation, a private company. In the latter half of this period, I concurrently held the posts of Chair of the Environmental Strategy Committee and the Technical Strategy Committee of the company and served as Chief Technical Officer, thus expending much energy on the question of how to balance the environment with economic performance in a corporation. I was, however, unable to come up with a satisfactory answer to this question. Not only did I have trouble finding the right answer; being unable to overcome the contradiction between environment and economy, I found myself being pulled ever deeper into the negative spiral created by this paradox. At the same time, I shuddered at the thought that without an answer to this question, manufacturing in Japan would have no viable future.

In order to search for an answer, I quit the company and took up a position at Tohoku University in 2004. Nine years have passed since then. Have I found an answer? Unfortunately, not yet, but I have started to get a feeling for where the doors are that must be opened to arrive at an answer.

On March 11, 2011, an earthquake of unprecedented scale struck the Tohoku Region where I work and live. If we think of civilization as the aggregation of technology, the dramatic way in which the cosmetic surface of civilization peeled off in an instant made me reconsider numerous issues. What does it mean to face the global environmental issue squarely? How can we contemplate the issue of living wholesome, fulfilling lives? And, what is the role of technology—now so urgently questioned—in contributing to these two issues? It struck me that finding solutions to these issues is the responsibility of the survivors of the horrible earthquake and also one way of paying respect to those who lost their lives. Here, indeed, lies the answer to the question of how to balance the environment and the economy.

One solution, undoubtedly, is to be found in a Nature Technology development system which intelligently harnesses the amazing powers of nature. In 2009, as the notion of "Nature Technology", if still vague, started taking shape, I published a book from Tohoku University Press entitled *Channeling the Forces of Nature*, with the help of two colleagues. In this book, I had not yet discovered the lifestyle approach which was one of the doors to answering the questions at hand, nor could I envision the actual methodologies required to design such lifestyles. Later, with the help of Dr. Ryuzo Furukawa, who has worked with me since 2005, and other colleagues, our research made significant progress, and the contours of a Nature Technology development systems started emerging. I feel as if the fog has rapidly started clearing. It is thus fair to say that the book you are holding in your hands is a reexamination of the philosophy of a Nature Technology development system and the actual approaches and methodologies involved.

The development of Nature Technology involves using a backcasting approach to envision lifestyles that are wholesome and fulfilling even under severe environmental constraints; identifying the technologies required to enable such lifestyles; searching in nature for the seeds of such technologies; and then—through the filter of sustainability—redesigning these as applicable technologies.

In our analysis of the lifestyles envisioned with backcasting, we found that many people strongly, yet often unconsciously, yearn for nature and enjoyment in their daily lives. We have also learned that in the development of technologies to enable such lifestyles, the notion of *iki*—described in detail in Chap. 10—which takes nature as its point of departure, is of the utmost importance.

There are still several hurdles we need to overcome in order to complete a system for the development of Nature Technology. It is, however, my sincere hope that the methodologies and research results presented in this book may help share new ways of thinking about how to shape exciting, wholesome, and fulfilling lifestyles even in the face of rapidly intensifying environmental constraints, thus also laying a few stones in the path to a new civilization based on a fresh set of values which do not merely extrapolate from the past.

Sendai, Japan   Emile H. Ishida
Okinoerabu Island of Amami Islands
2013.08

Acknowledgements

Is it possible to arrive at a balance between the environment and the economy? In this book, we have pulled together the outcome of our pursuit over the last 10 years of an answer to that question.

In September 2004, I initiated research at the Graduate School of Environmental Studies at Tohoku University on the issue of how to balance the environment and the economy from the viewpoint of Nature Technology. Regarding worldviews and civilization, I gained much insight participating as a research fellow in the Twenty-First Century Environment, Economy and Civilization Project headed by Professor Yoshinori Yasuda (presently professor emeritus at the International Research Center for Japanese Studies). In 2005, Dr. Ryuzo Furukawa (presently associate professor at the Graduate School of Environmental Studies, Tohoku University) started participating in my research—first as a researcher from a private think tank. Dr. Furukawa has actively pursued new research methods such as lifestyle design based on backcasting or interviews with people in their nineties, and he plays a central role in the lifestyle research which forms the backbone of Nature Technology development. In 2009, we were given the opportunity to set up a Nature Technology Research Consortium as part of the Japan Manufacturing Conference (MONODZUKURI Nippon Conference), which, with Nikkan Kogyo Shimbun as secretariat, enjoys the participation of more than 2,000 corporations. In this forum, we have been able to discuss the relationship between new technologies and ways of living and to promote concrete, new business initiatives. Meanwhile, the SEMSaT Course (Strategic Environmental Management and Sustainable Technology Solution), which I initiated in 2005 at the Graduate School for Environmental Studies—a course which mainly targets people already in work—allowed us to embark on numerous trials of the actual creation and implementation of business systems based on the notion of Nature Technology in the context of corporations, government organizations, and non-governmental organizations. While undertaking such activities, we received many valuable insights and inspirations from people in corporations that helped drive our research forward. Furthermore, in 2012 I had the opportunity to participate as a representative of researchers in the Innovative Materials Engineering Based on Biological Diversity Project (a new academic field under the auspices of the Ministry of Education, Culture, Sports, Science and Technology) initiated by Professor Masatsugu Shimomura (Professor, Advanced Institute for Materials Research, Tohoku University)—one of Japan's foremost biomimetics researchers—and thus was able to experience biomimetics research firsthand. At the same time, I was able to gain many valuable ideas and opinions on the new methodologies required to implement Nature Technology in society, and several concrete joint research projects were commenced. I wish to sincerely thank the many friends who were so kind as to participate in these activities.

I also wish to express my gratitude to the team of people who supported our efforts on a day-to-day basis: Dr. Ryuzo Furukawa, Dr. Yuko Suto, Dr. Hirotaka Maeda (presently at the Nagoya Institute of Technology), research assistant Koichi Okada, research assistant Tomoko Monobe, research assistant Noriko Konno, and the many students who have been affiliated with my research lab. I am most grateful for their dedicated and tireless efforts.

I also wish to thank my friend Peter David Pedersen, who provided much advice as we were setting up SEMSaT and who supervised the translation into English. I also thank Springer Japan for giving me the opportunity to publish this work. Without the understanding and great effort of the staff at Springer, it would not have been possible to share the findings of this book with a wider, international audience.

Finally, I wish to express my gratitude to my partner, Ako, who always takes great care to create the best possible environment for me to work in.

Sendai, Japan   Emile H. Ishida

Contents

Chapter 1
Where Are We Heading?

Keywords Backcast thinking • Forecast thinking • Global environment • Great East Japan Earthquake • Material richness • Nature technology • Spiritual richness

1.1 What the Great East Japan Earthquake Taught Us

At 14:46 on March 11, 2011, a magnitude 9.0 earthquake—the largest ever recorded in Japan—struck the Tohoku Region. The epicenter was 24 km under the sea off the east coast of Honshu at 38.1° North latitude, 142.9° East longitude. Seismic faults were activated in a vast area stretching 500 km north–south from the coast off Iwate Prefecture to Ibaragi Prefecture and 200 km east–west. The tsunami triggered by the earthquake caused immense destruction up along the Sanriku coastline, and the number of dead or missing reached approximately 20,000. In addition to this, a nuclear power plant (Fukushima Dai-Ichi), the showpiece of Japan's energy policy, was destroyed and damage caused that will require remedial action for decades if not more.

In some places, the tsunami exceeded 10 m and ended up destroying more than 22,000 ships and 300 fishing harbours, with some 23,600 hectares of farmland being laid to waste.

Gas, gasoline, oil, electricity and other energy supply systems were devastated, and transportation and communications infrastructure as well as water and sewage lifelines severed. Furthermore, Japan's most abundant food supply region suffered immense damage. In Japan, there are five prefectures (out of a total of 47) with a food sufficiency of more than 100 %, four of these located in the Tohoku region. Iwate, Miyagi and Fukushima Prefectures alone supply 13.1 % of the rice, 5.1 % of the vegetables, and 9.2 % of the meat and dairy products made in Japan. As for fisheries, also heavily damaged by the disaster, the Tohoku region accounts for 15 % (monetary value) of the national total (Figs. 1.1, 1.2, 1.3, and 1.4).

In "Channeling the Forces of Nature", which we published in 2009, we pointed out that if humanity continues with a business-as-usual scenario, environmental risk

E.H. Ishida and R. Furukawa, *Nature Technology: Creating a Fresh Approach to Technology and Lifestyle*, DOI 10.1007/978-4-431-54613-9_1, © Springer Japan 2013

Fig. 1.1 Houses carried inland to elevated areas by the tsunami (Onagawa)

Fig. 1.2 Most buildings along the coastline were completely destroyed (Kesen-numa)

will reach a critical threshold by around 2030, and that we may, thus, risk triggering the collapse of civilization. The devastation of energy supply systems and the break-down of basic infrastructure, as well as the severe damage to food production capacity experienced in the aftermath of the 2011 earthquake and tsunami, almost felt as a precursor to 2030. 20 years early, it were as if we have been given a preview of what

Fig. 1.3 A vessel near completion, destroyed and thrown up on the dock (Kesen-numa)

Fig. 1.4 A train in operation was derailed by the tsunami and broken into two (Kesen-numa)

2030 may hold. The technologies and approaches applied to overcome the effects of this disaster are thus very similar to the ones needed to tackle the global environmental problems likely to face humanity around 2030. There is a need today to join intellectual forces, to pool our knowledge, in order to come up with truly effective solutions.

The Great East Japan Earthquake caused modern civilization, in which we had such confidence, to shatter in an instant. Amid all this despair, however, it was also possible to detect some rays of hope. The hope was found in the attitudes, tools and art of living of the disaster struck people, who, even when they were fed only one rice ball in a day, stayed calm and, with a smile on their faces, were able to say "thank you." The calamity poses important questions that we need to answer. How can we deal effectively with the global environmental problem? What does it mean to live with a sense of fulfillment? And how may technology contribute to these two issues? Answering these questions is, we believe, our responsibility towards future generations.

1.2   Contemplating the Nature of Prosperity

What, in the first place, is the role of technology (and of government, corporations, NGOs)? It is, undoubtedly, to enable people to prosper and live satisfactory lives. It is no exaggeration to say that there is no other ultimate purpose. The important point here is that most people yearn for what we might call prosperity, or richness, of the heart as opposed to prosperity of things. In Japan, there has been a clear trend as early as the 1980s to value prosperity of the heart (the non-materialistic aspect) higher than prosperity of things (the materialistic aspect) (Fig. 1.5).

The question is, are technology, corporations, and government contributing to this sense of true prosperity? Unfortunately, although GDP has risen, happiness and life satisfaction levels have not followed suit, on the contrary, reality is a slow decline in the latter numbers (Fig. 1.6). In an opinion survey of 23 countries

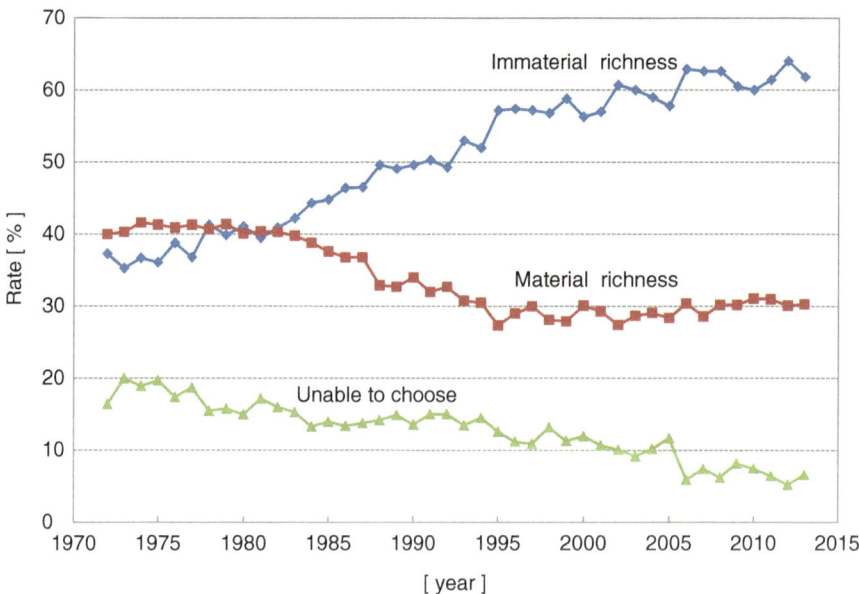

Fig. 1.5 Importance of immaterial richness versus material richness. Created on the basis of "Public opinion survey concerning people's lifestyles" (The Cabinet Office, Japan, 2013)

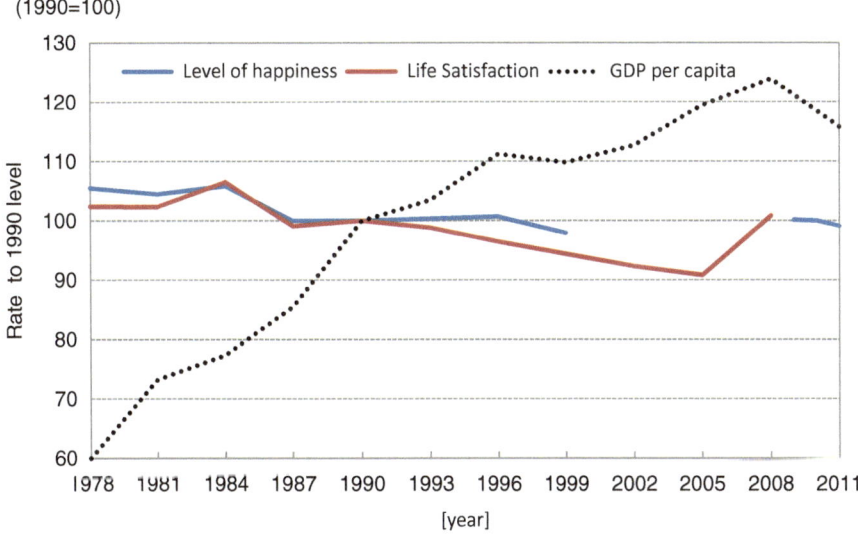

Fig. 1.6 Survey on perception of levels of happiness and life satisfaction. (*Source*: Research Committee on Happiness Levels, Aug. 2011, Japan)

(IPSOS2010), 86 % of Japanese respondents expressed anxiety about the future. A similar trend can, generally, be observed also in other industrialized countries, but still the figure for Japan is conspicuously high compared with the 79 % of French, 56 % of American, and 55 % of British respondents expressing the same concern. In another survey (Japanese Trade Union Confederation, 2010), only 8.5 % of Japanese responded that they thought Japan would be in a better state 1 year into the future, 32.6 % for 5 years, and 46.5 % when looking 10 years ahead, still less than half the population. Suicide rates in Japan increased dramatically in 1997–1998 and have stayed above 30,000 incidents per year. In 2012, the number finally dropped to 28,000, but the number of suicides among people in their twenties has not fallen. Judging from such data, it seems that among the developed nations, Japan is particularly caught in a mood of hopelessness. The Great East Japan Earthquake, as far as we can judge, only added to this sentiment. Why is the situation like this in Japan?

After the Meiji Restoration (1868), Japan for decades tried first to catch up with Europe, and then, after World War II, with the United States. The nation and its people were united in a frantic effort to grow and finally reached the peak of prosperity. This peak was, without a doubt, one of material prosperity with Japan arriving at the zenith of modern, materialistic civilization. Having arrived at that point trying to gain even further prosperity, did the Japanese maybe, at last, realize that were in a dense fog, unable to gain a view of the road ahead? We all know, deep down, that there can be no extension of the material prosperity gained in the past by exploiting a seemingly limitless supply of energy and resources. If we were to continue along

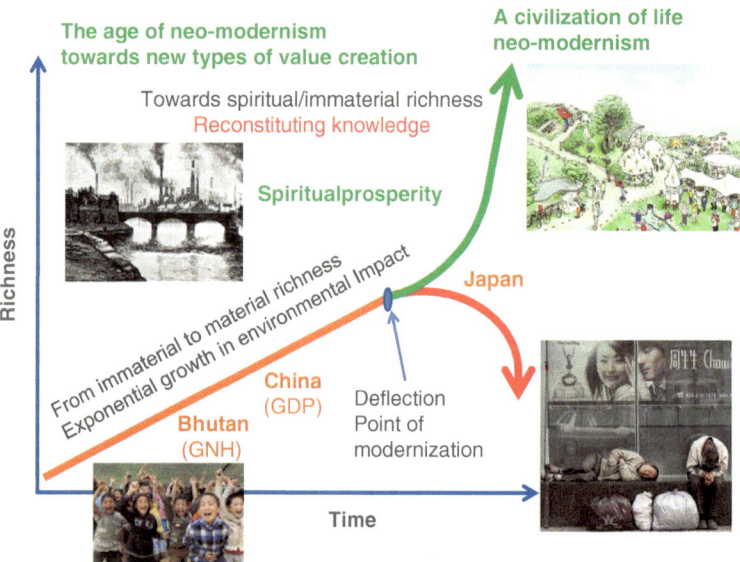

Fig. 1.7 Where should we go? Where is the next growth potential?

that road, it surely would mean that we, ourselves, would trigger the collapse of modern civilization (Fig. 1.7).

So, where should we then be heading? Neither the government, nor corporations or educational institutions have been able to set a new vision for the future. Trapped in past experiences of success, politics has gone astray, major corporations log huge deficits, and disillusioned students in ever larger numbers turn away from the study of science and technology.

We need to clear the fog and set a new direction for the future. That is, indeed, the very purpose of this book. At least, it is quite clear that this new direction cannot be an extension of the road travelled so far: there is a need to establish a new point of departure—a new platform for action. Our measurement can no longer be simply monetary; a new platform for action requires new measures. For example, why does an eco-car require 1.5 tons of materials and equipment to carry and individual weighing a mere 60–70 kg? Why, when we are told that even the weight of your golf bag left in the boot of your car will lower fuel efficiency, does an automobile need to carry several hundred kilograms worth of safety equipment? Why is it that people are still attracted by watches wound by hand, when we have high-tech, accurate quartz and automatic watches? Why are people, despite anti-ageing being the buzz-word of our age, still attracted by the deep furrows and wrinkles in the face of an old, sun-tanned fisherman? Why are the smiles of people in the primary industries, who often earn only a fraction of those in secondary or tertiary industries, so much richer? And, why is it that no waste is generated in the natural world? Today, we are more than ever in need of a new set of measures that would make it possible to respond to such questions.

1.3 Where Are We Heading?

Few people would disagree with the statement that the global environmental problem is an urgent task for humanity to deal with. Reality, however, is that the eco-technologies supposedly beneficial to the environment have ended up becoming like indulgences for further consumption, thus fuelling environmental degradation—a paradox we call the eco-dilemma. Eco-technologies have served as a wake-up call to conventional technologies which use a seemingly endless supply of resources to provide comfort and convenience, and the diffusion of eco-technologies has helped nurture a high environmental awareness in approximately 90 % of the Japanese. What they have not achieved, however, is to halt the deterioration of the natural environment. Eco-technologies are but one step in an evolutionary process and now, to complete this process, further creative destruction is needed. A creative destruction that helps reshape lifestyles is what we must pursue today (see Chap. 2).

Needless to say, to embark on this task a correct understanding of global environmental constraints is essential. Facing us today are seven different environmental risks: the depletion of energy and resources, the deterioration of biodiversity, the distribution of water and food, the rapidly increasing global population, and climate change most often experienced as global warming. These risks are intensifying at great speed and may well reach a critical threshold by around 2030. What, in essence, is the global environmental problem? Obviously, these seven risks are all very important, but did they also exist as risks 100 years ago? The answer is no. It is within the last 50 years or so that these risks have started to become a real threat to the continuation of our civilization. Why did these risks materialize? The answer, undoubtedly, lies mainly in the swelling of human activity. In order to obtain increasingly marginal improvements in comfort and convenience, we have continued an exponential growth of consumption and led ourselves into a cul-de-sac in which, by 2030, we may well experience a civilizational crash. The global environmental problem derives entirely from the endless expansion of human activity, and the response that we must now muster is how to cease or reduce the swelling of our activities, while at the same time realizing fulfilling lives, the essence of human existence. Our success will depend on whether we can make the necessary lifestyle changes (see Chap. 3).

Where can we find the answer to the great challenges we face? In order to realize a sustainable society, we must acknowledge the need for a resource-cyclical society with respect for the environment while at the same time accepting the irreversibility of people's perception of quality of life (by which we mean the nature of human desire—once people have reached a certain level of comfort and convenience, it is extremely difficult to accept letting go of this). We must clarify what is meant by lifestyles of true satisfaction—richness of the heart—considering the environmental constraints existing, and then explore what kind of technologies are required to make such lifestyles possible (see Chap. 4).

The question then is, will we be able to change lifestyles? Unfortunately, this is not an easy task to accomplish. Our way of thinking is dominated by a forecasting mentality which extrapolates from the present. If we use this mentality to think of the future of lifestyles, we must find solution to the question of how to achieve a prosperous lifestyle on the basis of presents levels of desire and protect the environment at the same time. Sadly, we cannot solve these two challenges simultaneously using conventional thinking. If we are truly concerned about the environment, we feel obliged to be frugal and hold back; if we choose to focus on living a materially prosperous life, we easily end up in a cycle of free-wheeling consumption. Whichever the choice, there will only be a partial solution to the problem, and as a result we end up creating eco-dilemmas. The new mentality required to arrive at a proper response, is thinking based on backcasting. This means focusing on what a wholesome, fulfilling lifestyle would entail in, for example, the year 2030, considering the severe environmental constraints we will be facing by then (see Chap. 5).

Using the backcasting method, the authors of this book have so far outlined more than 1,500 different lifestyles. Analyzing the social acceptability of these possible lifestyles, we have arrived at the conclusion that most people between 20 and 70 years of age, often without being conscious thereof, strongly seek "convenience" "enjoyment" "nature" "self-growth" and "a sense of belonging (to society)" in their lives. Surprisingly, we found that the degree to which people seek enjoyment and nature differs little from such a trait as convenience. But what is enjoyment? At present, people have easy access to electronic games, tv-sets, the internet and numerous other tools of enjoyment, but what people seek deep down are other forms of enjoyment. Or, what does nature mean in this context? That people want to go to the sea or hiking in the mountains? That is probably not the entire story. If we can uncover the true nature of such desires as enjoyment and nature in people's lives, we should find it easier to envision what truly prosperous lifestyles might look like (see Chap. 6).

To clarify what people mean when they emphasize the value of nature or enjoyment in their daily lives, we conducted a series of interviews with Japanese around the age of ninety (nonagenarians), and from this made a qualitative analysis of pre-WWII lifestyles, a time when people lived in closer proximity to nature. We discovered as many as 70 different types of wisdom or techniques that these people had used to live in a harmonious relationship with nature. The common, fundamental characteristics identified in this exercise were "acknowledging that humans are given life by nature", "making the best use of nature", and "skillfully dealing with the challenges put forth by nature". As we visited several disaster shelters after the Great East Japan Earthquake, we became convinced that these forms of wisdom or techniques of life constitute the very essence of living mindfully, and are elements the Japanese must not lose (see Chap. 7).

We thus found that the "nature" and "enjoyment" seen as attractive by most Japanese today were very vibrant elements of pre-WWII living. Nature, enjoyment and a sense of belonging (social communion) were all vividly present in the lifestyles enjoyed before the war. One could speculate whether it would be possible to return to such lifestyles, but unfortunately our desires and definitions of quality of life tend to be irreversible. More than 2 years after the 2011 earthquake and tsunami, the images of destruction remain sharp in our minds, and we know it will take considerable time for the wounds to heal. Our daily actions, however, already diverge from this mentality. A mere 100 days or so after the disaster, the initially experienced material frugality had more or less vanished, and as more time passed there even seemed to be a rebound effect. This happened despite the fact that we had experienced the interruption of electricity and gas supplies and even had trouble securing sufficient food. Changing lifestyles while acknowledging the existence of an "irreversibility of (notions of) quality of life", means that we must recreate the magnificent wisdom of living gained from our interviews with nonagenarians in such a way that it can be accepted by present generations. What are the criteria by which we might do this? Combining the analysis of social acceptability of different lifestyles based on backcasting with the precious keywords gained from interviews with nonagenarians, we have, finally, started to uncover the shape of wholesome, fulfilling living. This shape is characterized by the way in which constraints can actually help people grow and become more creative (see Chap. 8).

The next question is how, in concrete terms, to facilitate the lifestyle innovation required. The first step in this process is to optimize the utilization of already existing technologies based on underground resources (mineral and energy resources mined from the Earth's crust). That underground resource-based technologies today accelerate environmental degradation is the result of partial solutions deriving from a forecasting mentality. Lifestyles in which total system optimization is achieved based on backcasting can be created and then diffused with the help of a number of already existing technologies. We have chosen to call these "transformation technologies" in that they help initiate the next step of technological evolution— the step toward nature technology which wisely utilizes the forces of nature. One example of a transformation technology is a storage battery, which with a capacity of as little as 1 kWh may reduce the energy consumption of a household by 50 % (see Chap. 9).

The second step is to part with underground resource-based technologies. Nature has an innate intelligence with its own ethical principles. Since the Earth took shape some 4.6 billion years ago, and most obviously in the 3.8 billion years since the birth of life, natural selection occurred over and over again, and nature in the process perfected a completely cyclical system driven entirely by solar power. From this we are able to learn about mechanisms, systems and even the social process of creative destruction (selection). Since the Earth Summit in 1992, at least the developed nations have made efforts to create a sustainable society, but the chasm between ideal and reality is wider than ever. Now is the time to use backcasting to envision future wholesome and fulfilling lifestyles in the light of emerging global environmental constraints, and then to identify the technologies needed to realize these lifestyles searching for hints and guidance in nature. We must, thus, redesign technology through "filters of sustainability" shaping an entirely new mould for technological development. This is the very definition of a new technological development system based on nature technology.

Pursuing nature technology, we need to keep one further important perspective in mind. Most technologies today are products of the eighteenth century industrial revolution initiated in the United Kingdom. The British industrial revolution succeeded, in essence, by parting with nature while giving birth to the idea of mass production/mass consumption, ultimately triggering the global environmental crisis (see Chap. 10).

Will nature technology, which does not set us apart from nature, truly be able to contribute to the establishment of a sustainable society? History may give us a hint. While the industrial revolution, based on the principle of humanity's separation from nature, flourished over most of the world, there was a people who held onto a particular view of nature, and who promoted their own form of industrial development—namely, the Japanese of the Edo Period (1603–1868). Whereas the British industrial revolution evolved as a capital intensive enterprise, the Japanese version was a labour intensive revolution. As a result of this, the Japanese did not move toward mass production/mass consumption, but rather created technologies enabling play and entertainment. If we can say that the British industrial revolution created technologies fuelling material desires, it may be fair to say that the Japanese ditto was an industrial revolution fuelling immaterial (spiritual) desires, leading to the notion in Edo society of *iki*, meaning "liveliness of spirit". In this sense, we can say that nature technology, with nature as its foundation, incorporates the concept of *iki* (see Chap. 11).

The system of nature technology development has already begun to generate some concrete, new technologies. For example, a tub bath that does not require water, an air conditioner without a power supply, surface materials that do not easily get stained, wind generators that revolve in even the lightest breeze, and household farms... These technologies have nature as their foundation, are simple and easy to understand, encourage communication, and generate affection in their users (see Chap. 12).

Obviously, there are still many problems that need to be solved with these new technologies. How do we deduct the required technologies from projected lifestyles? How can we find the needed hints for development in nature? And, how can we actually redesign so that the technologies do not place a burden on the environment? These questions beckon answers, but we must, nevertheless, urgently take the first important steps. While thinking of the global environment, we ourselves must, with enthusiasm, personify the attraction of wholesome, fulfilling lifestyles and pass this spirit and experience on to the next generation.

Bibliography

Demura M (2011) Higashi-nihon daishinsai ni yoru suisangyouhigai to fukkou ni muketa kadai (Damage to fisheries and issues towards reconstruction caused by the Great East Japan Earthquake). Nourin Kinyuu 64(8): 489–503

Ichise Y (2011) Higashi-nihon daishinsai ni yoru nougyouhigai to kadai (Damage to agriculture and issues caused by the Great East Japan Earthquake). Nourin Kinyuu 64(8): 42–54

Ishida EH (2009) Channeling the forces of nature. Tohoku University Press, Sendai

National survey of lifestyle preferences, The Cabinet Office, Japan 1990

Public opinion survey concerning people's lifestyle, The Cabinet Office, Japan 2013

Quarterly estimates of GDP, The Cabinet Office, Japan 1990

Survey on perception of levels of happiness and life satisfaction, Research committee on happiness levels Aug. 20, Japan 2011

Chapter 2
The Eco-Dilemma

Abstract Although state of the art eco-technologies have been launched one after another, and many citizens have a high level of environmental awareness, the degradation of the global environment progresses steadily (the eco-dilemma). Why do we find ourselves in this situation? The reason is that eco-technologies have become like excuses—or indulgences—for consuming more, thus giving birth to a level of consumption which more than cancels out the positive contributions of technology. Part of the responsibility also lies with corporations that do not explain how properly to use eco-technologies. Are eco-technologies then an evil? It is true that eco-technologies have not been able to contribute greatly to preventing environmental deterioration, but they have surely helped nurture a high level of environmental awareness in about 90 % of Japanese citizens. What we have to consider today is how to continue a process of creative destruction to select the appropriate technologies. Today, after an era in which we pursued comfort and convenience on the basis of the insatiable consumption of resource and energy, we are experiencing the first step in this process of selection—a step which gave birth to eco-technology. But we must not rest on our laurels; the further step in selection needed is one towards the innovation of lifestyles. Today, we must make the transition to a new era in which technology assumes the appropriate responsibility for and in the creation of lifestyles.

Keywords Corporate brands • Eco-dilemma • Excuses for consumption • The coal problem • The uniformization of technology

2.1 The Global Environmental Problem and Our Way of Living

In 1992, 172 countries—more than 100 thereof personally represented by heads of state—gathered in Rio de Janeiro (United Nations Conference on Environment and Development), and jointly committed to make the greatest possible effort to create

E.H. Ishida and R. Furukawa, *Nature Technology: Creating a Fresh Approach to Technology and Lifestyle*, DOI 10.1007/978-4-431-54613-9_2, © Springer Japan 2013

a sustainable society. Already 20 years have passed since that event, and while many countries are indeed making significant efforts, reality is far removed from the ideals and goals discussed at Rio, and, sadly, the degradation of the global environment is accelerating. Why are our efforts not paying off? The time has come to return to basics and reconsider what must be done.

2.2   The Nature of the Eco-dilemma

In Japan, it is no exaggeration to say that almost all aspects of society have, recently, become eco-focused. In the business world, a company can hardly market a new product any longer unless it takes environmental issues into consideration. Needless to say, some of these products belong to the "green wash" category (products that merely pretend to be green), but at least in Japan, such products are not accepted in the market, and corporations truly are making highly noteworthy efforts.

The relationship between energy consumption and electrical appliances in the household may serve as a good example. In 1965, electricity consumption accounted for 22.8 % of total energy consumption in the household; in 2008 the figure was 50.1 % (in this period, population increased by a factor of 1.29 and household energy consumption by a factor of 2.2). Of household electricity consumption in 2008, air conditioners accounted for 25.2 %, and refrigerators for 16.1 %—that is, more than 40 % of the total for just these two appliances. On the other hand, in the last 15 years the efficiency of air conditioners has improved by 40 % and that of refrigerators by an impressive 80 %. In other words, a refrigerator today can operate on only 20 % of the energy used for the same size appliance just 15 years ago. These improvements are the fruit of world class technological development by Japanese corporations.

The stand-by power for electrical appliances, which some years back drew considerable attention, has also decreased by 71 % in the period from 1999 to 2008, and now is estimated to account for about 6 % of total household energy use. And this is not the end of the story—TV-sets, LED lighting and automobiles are further frontiers of eco-technology. For automobiles in particular, the hybrid car triggered an epochal shift towards eco-cars, in which even light motor engine cars (below 660 cc. gasoline engine) today aim to be green. Products across virtually all categories have become eco-focused, and in a country like Japan where citizen awareness is high, there are virtually no fake "green-wash" eco-products to be found on the market. Japan has managed to create and market a continuous stream of highly advanced eco-technologies to an extent not seen in any other country in the world. At the same time, the environmental awareness of citizens is by far the highest of any industrial country. Our own surveys reveal that approximately 90 % of people are concerned about environmental issues, and 70 % are already taking action or consider it imperative to do so.

If this synergy between the launch of eco-products and high citizen awareness were truly effective, environmental trends in Japan should have improved

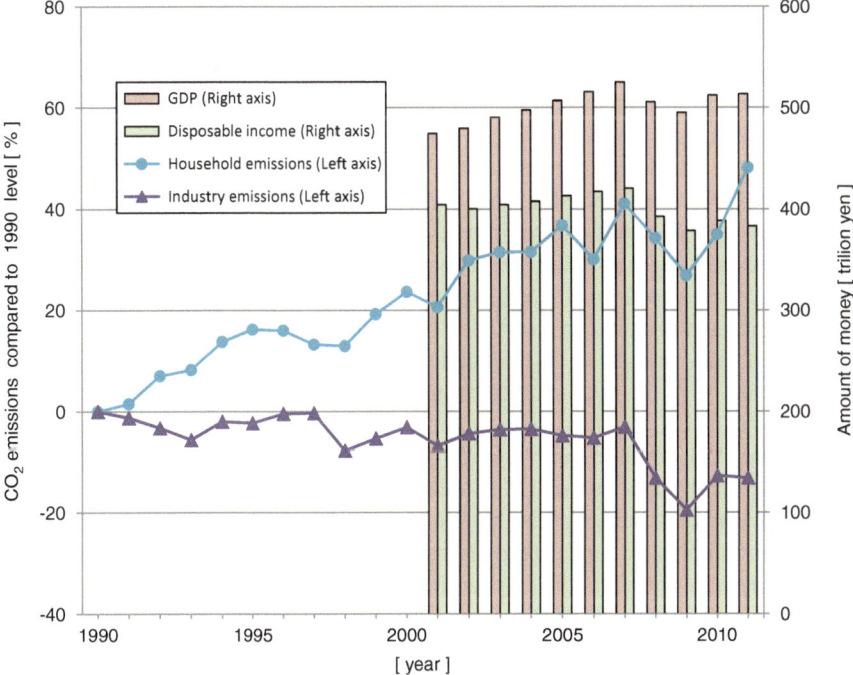

Fig. 2.1 Increasing environmental impact (CO_2) in Japan. Created on the basis of "National Account of Japan" (Cabinet Office 2013), "Family Income and Expenditure Survey" (Ministry of Internal Affairs and Communications 2012), and (Greenhouse Gas Inventory Office 2013)

drastically, but reality, unfortunately, is quite the opposite. The environmental impact (CO_2 emissions) of households, for example, continues to increase and in 2010 reached an abominable 130 % of the 1990 level (Fig. 2.1). In Japan, one factor of importance is that, despite a decline in population, the number of households has increased, but even considering this, the deterioration of the environment has undoubtedly progressed. Why does environmental deterioration continue even in a situation where innovative eco-products reach the market and general awareness is high? This is what we call "the eco-dilemma."

It is crucial today that we understand the nature of this eco-dilemma and, based on this understanding, make the necessary course correction. Why does this dilemma occur? Unless we understand the nature of the eco-dilemma, technologies introduced to the market in good faith may be of no use whatsoever—or even worse, they may fuel the fire of environmental destruction. In our surveys, we have discovered that the eco-dilemma fundamentally derives from two different factors. One is that, apparently, eco-products have become like excuses for consumption. People are easily tempted to consume more. When air conditioners become more efficient (and thus consume less electricity), people will buy another one for the bedroom. When the TV-set becomes more energy efficient, more and more people will buy larger sets. And when they buy an eco-car, and thus need less petrol, the distance of travel tends to go up. Encouraging this kind of consumption were government measures, finally discontinued in June 2011, allowing people to drive ad libitum on the highway during weekends for just 1,000 yen, and the eco-point scheme for electrical appliances (which awarded the consumer more eco-points the larger the size of the appliance purchased, as long as it belonged to the energy efficient category). Since the government promoted these schemes, it was easy for the general public to be lured into thinking that the more (eco)-consumption the better, and the paradoxical result has been that environmental impact is increasing. In a way, the government has issued official indulgences allowing citizens to feel comfortable consuming more.

The second factor in the eco-dilemma is the fact that the proper usage of eco-technologies and products is insufficiently explained by manufacturers. In fact it is worse than that—in some cases one might almost get the impression from operation manuals that since the product is environmentally oriented, the purchaser need not be concerned about how heavily it is used.

In Japan, 25.19 million flat screen TV-sets were sold in 2010, a whopping 185 % of the figure in the previous year. 8.24 million air conditioners were sold in the same year, both numbers being the highest in history. Considering that there about 50 million households in Japan, this means that on average every other household bought a new TV-set within just 1 year. Households with only a single occupant are increasing steadily, further fueling a growth in consumption (in 2010, the average number of household residents was 2.47). Already, there is an average of more than 2.4 TV-sets per household, for air conditioners the number has reached 2.5/household, and even for refrigerators the average today is 1.3/household. The notion that if only eco-appliances are bought by all households environmental impact would fall, is clearly an illusion (Fig. 2.2). Energy consumption has not fallen—on the contrary, it keeps rising. Even if the energy consumption of each individual appliance drops, the purchase of a larger number of units for each household combined with an increase in the number of households means that environmental impact, far from decreasing, continues its unabated increase. Obviously, an increase in the number of households leads to a rise in the number of refrigerators and TV-sets, as well as other items and appliances used in everyday life.

Even as this trend continues, 40 % of the Japanese say that they feel stressed by the excessive ownership of things and/or by being surrounded with too many devices and gadgets. TV-sets and refrigerators have become more efficient, but at the same time ever larger in size; air conditioners conserve more energy, but now adorn the walls of all rooms in the house; family size decreases, but the number of households continues to rise…

It is of course essential that each product and service becomes ecologically sound, but at the same time, we cannot ignore the importance of how they are used and what kind of lifestyles people pursue. Looking at the consumption issue in a global perspective, something similar is happening with automobiles. Fuel efficiency is improving year after year, but the rapidly increasing number of cars on the road more than cancels out the effect of this. Looking back at the 50 years from 1950 to 2000, world population increased by a factor of 2.4 from 2.52 billion to 6.06 billion, while the number of registered automobiles grew by a factor of more than

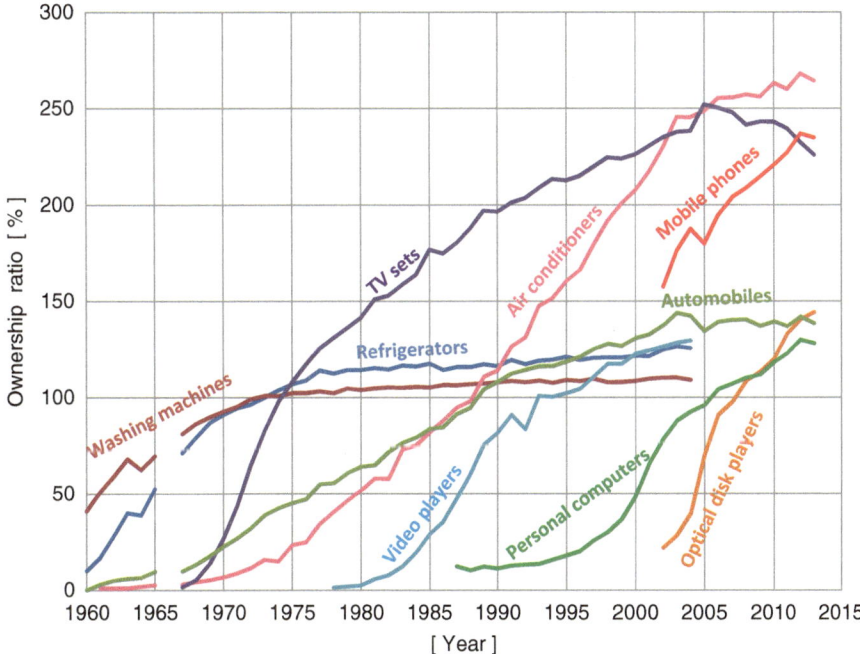

Fig. 2.2 Electric appliances in the home ownership ratio per household. Created on the basis of "Trends in Household Consumption" (Cabinet Office 2004, 2013)

10 from 70 million to 720 million, oil consumption by a factor of seven, and electricity consumption by a factor of 21.

If this trend continues, we may develop any number of new technologies without seeing a contribution to the reduction of overall environmental impact. In the long term, this may well mean that technology itself will be seen as an evil.

Concern about this eco-dilemma is not a new phenomenon. In 1865, William S. Jevons wrote a book entitled "The Coal Question". At the time, people who were concerned about the depletion of coal reserves due to the diffusion of the steam engine, argued that the development of more efficient engines would help alleviate this problem. Jevons, however, objected to this view. Certainly, he said, if you look at each individual steam engine, coal consumption will drop. The problem was that more efficient engines would become more convenient and economical to use, and thus be more widely adopted in society, in turn leading to a situation in which the increase in efficiency is more than cancelled out by the growing number of engines in use. In some cases, what is beneficial in a micro-perspective, may not be so when we look at the macro-perspective.

Enhanced efficiency in individual devices may well trigger a rebound effect in which more widespread usage causes overall environmental impact in society to increase. The discussion concerning Information and Communications Technology (ICT) in the 1990s is of a similar nature. The so-called Constant Time Budget

Hypothesis argued that the average time an individual traveled/moved per day is about 75 min, regardless of culture or level of economic development. People on low incomes may have no choice but to walk, while people on higher incomes use automobiles or airplanes to travel. The intriguing discovery, however, was that, regardless of the means of transportation, average time of travel/movement per day was approximately 75 min. The key factor here is not distance, but time. Following this theory, even if telework (powered by ICT) allows people to reduce their time of travel to and from work, the extra time gained will merely be used for other travel purposes. As a result, also, of the increased speed of travel made possible by less congestion, total distance traveled will actually increase, leading to a corresponding rise in CO_2 emissions. This is the rebound effect that may well occur. For corporate leaders, efficiency may have risen, but the evolution of ICT has not contributed to a decrease in environmental impact. On the contrary, it has made employees busier than ever, and the supposed increase in time available for leisure does not appear likely to materialize.

Let us return to the main story—why is it important today to refocus on the eco-dilemma? The reason is that the dilemma derives from the finite nature of the global environment. All things upon which the survival of humanity depends, from resources and energy to biodiversity, are finite, and we know that the risk of their depletion is increasing day by day. In this situation, we can no longer tolerate the existence of the eco-dilemma, and if technological development contributes to the aggravation of the dilemma, we are ourselves about to trigger severe environmental destruction and the collapse of modern civilization.

So, does that mean that the eco-dilemma as such is of a malign nature? We do not believe that is the case. Eco-technologies were born from a sincere reflection on the way we have been burdening the environment through the limitless use of resources and energy driven by our single-minded pursuit of comfort and convenience. Such a shift in thinking has also taken place in the minds of citizens—in Japan, for example, as mentioned above, close to 90 % of people already show a high level of environmental awareness. This shift is indeed very significant, but technologies that simply substitute one for another end up becoming excuses for consumption without contributing to a decrease in environmental impact. According to the Nikkei Newspaper (Oct. 12, 2012), the reduction of CO_2 achieved through the eco-point scheme was a mere one thirteenth of the initial projections made by the Ministry of Environment. This figure, one might say, is quite emblematic of the situation caused by the eco-dilemma.

What, then, is to be done? If partial optimization (as in the case of increased efficiency of individual appliances) ends up leading to an increase in the overall deterioration of the environment, we will have to reconsider what total system optimization (total-system improvement) means. Eco-technology is only one step on the way forward; we now need to move on to the next stage of creative destruction. This evolutionary step involves the creation of new, innovative lifestyles (Fig. 2.3). From the viewpoint of manufacturing, our range of choices is extremely limited. Technology must assume responsibility for lifestyles, and that, in turn, means that corporations must make clear what kind of lifestyles they aim to promote and launch

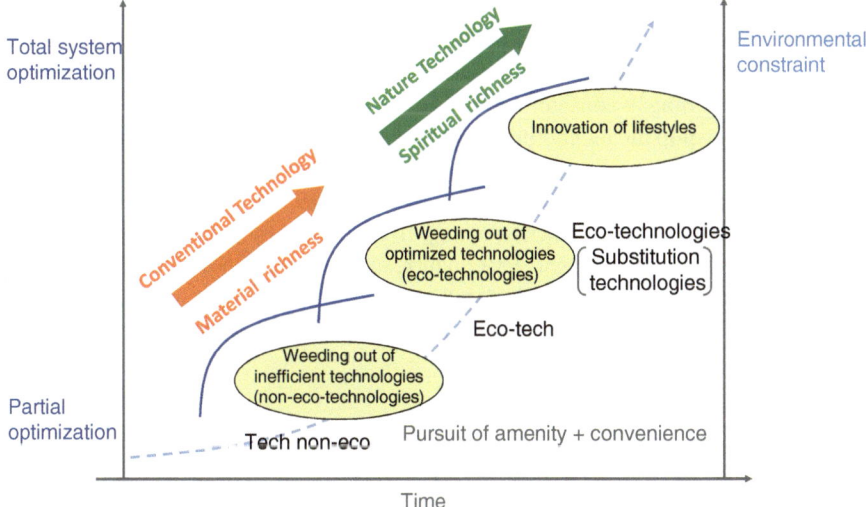

Fig. 2.3 Further weeding out, beyond eco-technologies, is required

technologies that help realize such lifestyles. So far, the main focus has been on how to use already existing technologies in society. As a result of this approach, human activities have swollen, thus causing the global environmental problem. Prosperity as such is necessary to enable decent, humane lives, and technology supposedly exists to help create this prosperity. Technology really is nothing more than a set of tools enabling human lives. Recently, however, we live in a technology-centered world in which most people believe they cannot live satisfactorily without great material wealth. We are thus in the process of creating a society in which technology, based on the supremacy of production, becomes the main actor. Of the Seven Deadly Sins Mahatma Gandhi warned the world of, we do indeed seem to be conducting "Commerce without Morality" and "Science without Humanity." We live in an era in which we seriously need to reconsider the significance of ethics and morality in manufacturing.

The difference between partial optimization and total system optimization in action taken to protect the environment, corresponds well to the difference between thinking based on forecasting and on backcasting, as we shall discuss in a later chapter.

While partially optimized eco-technologies have aimed for a balance between the provision of prosperous lifestyles and the protection of the environment, the end result has been the creation of eco-dilemmas. The biggest problem here is that the definition of a "prosperous lifestyle" is vague, allowing eco-technologies to be developed on the basis of the idea that "people's desires=convenience and comfort=a prosperous life". As long as material desires and convenience serve as the basis for action, achieving a balance between the provision of prosperous lifestyles and protection of the natural environment will be extremely difficult. If we look at the same issue from the perspective of total system optimization, it becomes clear that the basis for action must be "the finite nature of the Earth's environment". We need to acknowledge the existence of unavoidable environmental constraints before we can seriously consider the issue of what constitutes prosperous lifestyles, or of how to develop the technologies to allow for their realization. From this angle, prosperity is not the same as convenience. In some cases, a certain degree of inconvenience requiring human engagement (instead of a technological fix) may indeed lead to the feeling of true prosperity, arising from, for example, the commitment and affection born in the engagement. That, in turn, means that technology must no longer merely pursue micro-level efficiency, but must aim to contribute to a new form of prosperity while assuming responsibility for the role it plays in lifestyle creation.

2.3   Eco-products May Lower Corporate Brand Value

For corporations, a larger issues appears on the horizon—that of the uniformization of technology. If the development of eco-products becomes the ultimate goal, all manufacturers are obliged to develop more or less the same type of eco-technologies. For example, technologies that are resource efficient, energy conserving, thin/flat and light-weight, or that use biodegradable plastic, etc. All manufacturers will rush towards the same kind of technology development targets, killing individuality and creativity in the process, while generating uniform lines of eco-technologies. This is what we call the uniformization of (eco-)technology. When this happens, the only frontier of competition becomes the price of a product, thus inviting a fierce struggle to lower prices. In this kind of business-as-usual scenario, corporations have to mass produce in order to lower costs and survive cutthroat price competition, and the end result is conventional manufacturing with a thin layer of eco-camouflage. Continuing along this path may initiate a vicious cycle in which corporate brands are adversely affected. The FY2012 business results of major Japanese electronics manufacturers, some of whom fell deeply into the red, display this unfortunate trend all to clearly. Why was nobody able to foresee what now seems so obvious?

Most likely, the automobile industry will soon move down a similar path. Based on the equation that a good fuel economy=an eco-car, the high-cost hybrid car with a combined petrol engine and electric motor has become the flagship of eco-cars. The hybrid car does indeed boast an advantage in terms of fuel economy, but since the end of 2011, even gasoline engine light motor vehicles with an excellent fuel economy of more than 30 km/L have been brought to market. Will the next battlefield for eco-cars also be cost and price? Why have there not been other forms of value creation with more diverse forms of measurement—like beginning with the question of whether one really needs 1.5 tons of equipment to move an individual weighing some 60–70 kg? Is it ecological, even in the name of safety, that a car must be burdened with several hundred kilograms of safety equipment? Invariably, all things are being looked at through conventional lenses, when, in reality, we need to think of how technology should evolve in relation to changing environmental constraints.

Who could have imagined just 10 years ago that in agriculture, to take but one example, urban citizens would be willing to actually pay to experience rice planting or harvesting? New approaches will not appear when forecasting in a straight line from the present; only when we backcast from a desired future will new worlds of opportunity be revealed.

Ecology, or environmental consideration, is not a final goal to aim for, it is the point of departure, the starting line. Ecology is the natural state of affairs and must function as the foundation for manufacturing and corporate management. Only with this foundation can we start building wholesome, fulfilling lifestyles for people. Looking at the nature of the eco-dilemma, it is clear that an era has arrived in which technology, as mentioned above, must assume responsibility for lifestyles.

The eco-dilemma may be so evident partly because of the existence of detailed statistics in an industrial country like Japan. In any case, however, it is hard to believe that this situation is unique to Japan. If, sooner or later, not only the industrialized countries but a much wider circle of nations experience the same structural problem, we believe Japan has a role to play as a representative of the industrialized nations in facing the eco-dilemma squarely. In our view, it is no exaggeration to say that we in Japan should take on the role of creating a new approach to and new values for manufacturing.

Bibliography

Francesco Nachira, Paolo Dini and Andres Nicolai (2007) A network of digital business ecosystem for Europe, Digital Ecosystems Org. 1–20

Electronics and Information Technology Industries Association (2011) Kadenseihin no hanbai-jisseki suii (Trends in actual sales of electric appliances). Electronics and Information Technology Industries Association, Japan

Energy Conservation Center (2010) Handbook of energy & economy statistics. Energy Conservation Center, Japan

Ministry of Economy, Trade and Industry (2010) Energy Whitepaper 2010. Agency for Natural Resources and Energy, Japan

National account of Japan, Cabinet Office 2013

Family income and expenditure survey, Ministry of Internal Affairs and Communications 2012

Ferguson N (2009) Manee no shinkashi (The ascent of money), Trans. Senna, Osamu, Hayakawa Shobo, Tokyo

Gesell S (2007) Jiyuchi to jiyuukahei ni yoru shizenntekikeizaichitsujo (Die natürliche Wirtschaftstordnung durch Freiland und Freigeld), Transl. Aida, Shinichi, Pal Shuppan, Tokyo

Ishida H, Furukawa R, Dentsu Grand Design Laboratory (2010) Kimi ga otona ni naru koroni. Kankyou mo hito mo yutaka ni suru kurashi no katachi (By the time you grow up – a way of living that makes both the environment and people rich). Nikkan Kougyou Shimbun, Tokyo

Labor and Welfare Statistics Association (2010) Nihon no setaisuu no shourai toukei (Future statistics of the number of Japanese households). Labor and Welfare Statistics Association, Japan

Peart S (2006) Jebonzu no keizaigaku (The Economics of W.S. Jevons), Trans. Ishida, Haruo; Sekiya, Kizaburou and Kurita, Zenkichi, Taga Shuppan, Tokyo

The Cabinet Office (2010) Consumer confidence survey. The Cabinet Office, Japan

Trends in household consumption, Cabinet Office 2004, 2013

Transition of carbon dioxide excretion, Greenhouse Gas Inventory Office 2013

Uzawa H, Uchibashi K (2010) Hajimatteiru mirai (The future that has already begun), Iwanami Shoten Publishers, Tokyo

Chapter 3
The True Nature of the Global Environmental Problem

Abstract Energy—we have clearly already passed peak oil; biodiversity—we are using 150 % of the reproductive capacity of the ecosystem; global warming—the dispersion of aerosols into the stratosphere, a measure that once initiated must be continued indefinitely, is being discussed as a realistic countermeasure. Have we not lost sight of the true nature of the global environmental problem? Excluding issues that relate to the social sciences, we can identify seven major environmental risks facing humanity today. They are the depletion of energy and resources, the deterioration of biodiversity, the distribution of water and food, the rapidly increasing global population, and climate change most often experienced as global warming. These risks are all extremely important, but were they also risks 100 years, or even 50 years ago? What has made them such severe risks that they may threaten the very continuation of civilization? That, indeed, is the real global environmental problem.

Today we must first acknowledge that the global environmental problem derives from the excessive expansion of human activity, and then mobilize our joint intellectual forces in order to find out how we may cease or reduce this swelling while at the same fulfilling that most fundamental of human desires—to live a wholesome, fulfilling life.

Keywords Aerosols • Climate change • Ecological footprint • Geo-engineering • Global warming • Peak oil • Seven risks • Shale gas • Shale oil • The global environmental problem • The swelling of human activities • Trade-off

3.1 Is Global Warming a Global Environmental Problem?

To deal effectively with the eco-dilemmas caused when eco-technologies, paradoxically, end up becoming excuses for further consumption, we must reexamine the true nature of the global environmental problem.

E.H. Ishida and R. Furukawa, *Nature Technology: Creating a Fresh Approach to Technology and Lifestyle*, DOI 10.1007/978-4-431-54613-9_3, © Springer Japan 2013

There are a vast number of angles from which one may consider the global environmental problem; to begin with let us try to sort these out. The question, for example, of how much longer the Earth will continue to rotate on its axis is indeed a global environmental problem. The issue of how long biodiversity will exist at present levels, or of how long a particular species will be able to survive can also be said to be important global environmental problems. The global environmental problem is obviously multi-facetted, but the key issue that we must consider today is how long the human species can continue to exist on Earth. From this angle, how may we understand, for example, global warming in relation to the global environmental problem? And, how may technology contribute to the solution of this problem? Here, we shall take as an example the issue of geo-engineering, which is one of the hottest topics in documents from various country delegations at scoping meetings for the 5th assessment report of the IPCC (Intergovernmental Panel on Climate Change), due in 2014 (For details, see Sugiyama 2011). Geo-engineering is defined as a large scale, deliberate modification of the natural environment undertaken to alleviate global warming. Historically, this approach has its roots in military research in the 1960s, when large budgets were allotted to the study of deliberate modification of the environment as a potential weapon in future wars. A top secret project, codenamed POPEYE, to cause artificial rain in southern Vietnam during the Vietnam War has been brought to light (also known as the Watergate of climate war). Technology to modify meteorological phenomena was actually used on the battlefield. A more recent example, fresh in memory, is that of China announcing in 2008 that it would control rainfall in order to keep skies clear at the opening ceremony of the Olympic Games (August 8th). The Chinese Secretariat for Artificially Influenced Weather fired 1,104 rockets loaded with silver iodide into clouds around Beijing to induce rain, and thus succeeded in evaporating all rain clouds by August 6th, before they reached Beijing. The dry air caused by this initiative generated cumulonimbus clouds over the East China Sea leading to heavy downpours and thunderstorms over the entire Japanese archipelago.

Geo-engineering to prevent global warming largely falls into two categories: one is technology to remove CO_2 from fuel or from the atmosphere, the other technology to block light from the sun before it reaches the Earth's surface. To the former category belongs efforts to artificially strengthen the CO_2 capturing capability of nature, for example through afforestation on land areas, through the dispersal of iron or other minerals to enhance the absorption capacity of the sea, or in a chemical process (CCS=Carbon Dioxide Capture and Storage) using calcium or magnesium silicate to capture CO_2 from the atmosphere. The latter category focuses on increasing the reflection rate of sunlight entering the Earth's atmosphere, for example by dispersing sulphur oxide in the stratosphere, increasing atmospheric aerosols, or stirring sea water in order to raise the salt percentage allowing for an increased formation of clouds over the area, thus raising the reflection rate. Other initiatives under consideration include establishing a 3 million km^2 sunlight shield in space about 150 km above the Earth, and setting up reflection sheets in deserts or on rooftops to change the reflection rate of the Earth's surface.

The reason serious attention is recently being paid to such initiatives is the utterly slow pace of greenhouse gas reductions. As the outcome of the UN Climate

Conference in 2009 in Copenhagen (COP15) shows, efforts to reduce the emission of greenhouse gases have virtually come to a halt. A further factor is the uncertainty found in the relationship between greenhouse gas reduction and the future alleviation of global warming. For example, scientists estimate that even if we reduce the emission of greenhouse gases by half by 2050, there is still a 12–45 % chance that the Earth's average temperature will rise by more than 2 °C. A temperature increase of more than 2° could trigger abrupt and dramatic climatic changes—the Amazon rainforest may rapidly diminish in size, the sea ice of the Arctic sea and the ice shelves of the Antarctic continent melt at an accelerated rate, and the methane hydrate contained in sea floor sediments be released adding to the greenhouse effect.

It is with these risks in mind that the interest in geo-engineering as an artificially induced way of reducing the Earth's temperature is growing recently. One of the approaches to which most attention is being paid is the injection of aerosols—fine solid particles or liquid droplets—into the stratosphere. There are several different types of aerosols, including, for example, sulphate aerosols, dust, or aluminum oxide. The effect of injecting sulphate aerosols into the stratosphere is known from naturally occurring volcanic eruptions. The 1991 eruption of the Pinatubo volcano in the Philippines, which emitted huge volumes of sulphate aerosols, led to a decrease of surface temperature by about 0.5 °C. Geo-engineering aims to do the same artificially.

This method has the potential to lower the Earth's temperature almost immediately, and can be undertaken at a very low cost. In the Stern Report (October 2006) on the economics of climate change, the cost of climate measures to realize a low carbon economy and society are estimated at 1–2 % of annual global GDP. This means that, annually, about one trillion dollars is required. Geo-engineering in the form of injection of aerosols into the stratosphere, however, would only cost around four billion dollars. On the other hand, there are also concerns regarding this approach. Sulphate aerosols destroy the ozone layer and reduce the amount of sunlight reaching the Earth's surface, which, in turn, may cause decreased evaporation of sea water and thus less rainfall. Dispersing aerosols not only reflects the sunlight, but also makes the sky turn white leading to blood red sunsets. Other possible consequences include the blurring of starlight and the obstruction of astronomical observation. Apparently, research in the United States has even begun to look at how people would psychologically respond to white skies and blood red sunsets. The ability to disperse aerosols into the stratosphere is not limited to developed nations; indeed, any country with space technology, or in extreme cases even individuals or corporations could theoretically do the same. To developing nations, for whom the future environmental burden seems certain to grow, this may appear to be a highly attractive approach. An even greater concern is that once you embark on an initiative like this, it must be continued indefinitely. If we choose to reduce global warming through the continuous dispersal of aerosols, then a discontinuation thereof would lead to rapid increases in the Earth's temperature. Another aspect to consider is what the impact of such a measure would be on the Earth's ecosystems. If less sunlight reaches the surface of the planet, fewer clouds will form, as mentioned above, and precipitation may fall. But this is not the only outcome. The photosynthesis of plants could be impaired exerting great influence not only on the supply of water, but also on food production. If ecosystems are heavily influenced this not only impacts biodiversity, but quite possibly also the health of humans. Just as in the case of genetic modification, the issue of geo-engineering inevitably requires a deep, ethical debate.

The founder of Microsoft, Bill Gates, is said to have invested some 4.5 million dollars in geo-engineering and to have already applied for several patents. Also, in Australia there, apparently, is a business venture dispersing iron on the sea to increase growth of plankton, thus raising their ability to absorb CO_2. The business model, we are told, is to sell the CO_2 reductions under the United Nations Clean Development Mechanism.

3.2 Technological Progress Involves Trade-Offs

How, then, should we evaluate this new technology called geo-engineering? Of course we should not disregard its potential value and research probably needs to continue. At the same time, we need to establish international guidelines with a sophisticated ethical foundation. Most important, however, is that we stop and reconsider the true nature of the global environmental problem. Most technologies in use today have been developed after the industrial revolution in the United Kingdom and are based on underground resources and energy (mined from the Earth's crust). It is an undeniable fact that present global environmental problems derive from our use of these underground resource-based technologies. Neither can we deny that the history of technological progress is a history of trade-offs. Technologies developed in good faith always have unintended side effects, and as soon as we find a way to remediate these, new problems occur. It is through this continuous process of trial and error that technology has developed. What some called the greatest invention of the twentieth century—odorless, harmless freon gas—turned out to be destructive of the ozone layer and to have a global warming potential 20,000 times higher than that of CO_2. As the Jevons paradox, described in Chap. 1, pointed out, an improvement in the efficiency of a technology may well lead to an increase in its use and therefore a rise in environmental impact. And, as we have learned with antibiotics—supposedly a miraculous wonder drug—their widespread use has led to the birth of highly resistant strains of bacteria. Although we are aware of this, we are still obliged to try developing even more powerful medicines.

Geo-engineering is an approach that tries to solve a problem created by humans—global warming—with an artificial (humanly conceived) modification of the Earth's environment. In particular, considering the case of altering the reflection rate of sunlight reaching the Earth by dispersing aerosols into the stratosphere, the authors of this book do not think this is the best way to be expending our wisdom. We should not initiate an approach such as this, which, once embarked upon, cannot be discontinued safely. And, we need to have an overall perspective on how such a technology might influence the ecosystem, the supply of water and food so vital to life, and the provision of energy and resources for manufacturing. We should try to identify the principles that enable the development of technologies with a minimum degree of trade-off, and then set about actually realizing these technologies. The moment we think any particular issue, such as global warming, is the only important global environmental problem, and thus focus our efforts single-mindedly on lowering temperatures, we find ourselves trapped in an inescapable trade-off that may place a severe burden on nature's ecosystem and, in turn, end up threatening the human lifesphere.

3.3   The Global Environmental Problem Is Rooted in the Bloating of Human Activity

Let us take another look at the true nature of the global environmental problem. Excluding issues relating to the social sciences, we can say that humanity is facing seven risks (for details, see "Channeling the Forces of Nature", published by this book's authors) (Fig. 3.1).

They are the depletion of energy and resources, the deterioration of biodiversity, the distribution of water and food, the rapidly increasing global population, and climate change most often experienced as global warming. If humanity continues along the present path, most of these risks will reach a peak around 2030, but even with advanced science making an all out effort, we are still unable to predict which one of the risks will be the most dangerous one. Whichever it may be, the probability that we will trigger the collapse of modern civilization by 2030 if we continue along

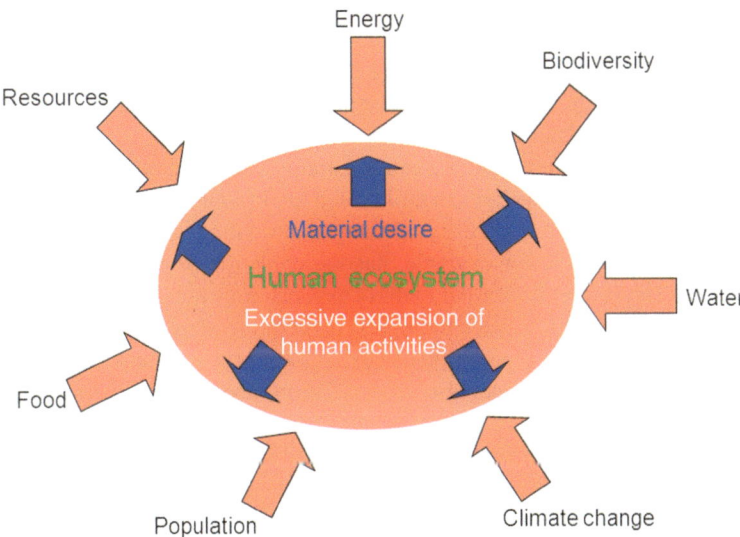

Fig. 3.1   The global environmental problem = the expansion of human activity

the present path is uncomfortably high. The authors of this book have been arguing that peak oil might arrive around the year 2030, but in 2010 the International Energy Agency (IEA) officially admitted in its annual World Energy Outlook 2010 that peak oil occurred in 2006. Predictions are that oil production will fall by a few percentage points every year, and although there may be price fluctuations in the short term, a continuous rise in prices seems certain for the long term. The possibility that our oil independent civilization will end much earlier than expected has increased. The New York Times on Nov. 24th, 2010 asked "Is 'Peak Oil' Behind Us?, and Le Monde of Nov. 12th the same year, in an article entitled "The End of Cheap Oil" wrote, "There is no time for denial. Government and communities need to start adapting now". Meanwhile, full scale production of shale gas and shale oil has begun the United States, leading to a drastic fall in the price natural gas, creating economic ripple effects such that the talk is of a "shale revolution". If, however, we start accelerating the extraction of these resources, what will there be left for us to hand over to future generations?

Looking at the issue of biodiversity, the Living Planet Report 2010 (WWF) reported a drop in the number of species in tropical regions of some 60 % between 1970 and 2007, and a fall of some 30 % for all types of ecosystems on Earth. The same report looks at the ecological footprint of human activities and concludes that whereas the available area for biological production on Earth is 1.8 global hectares (gha)/person, our actual footprint is already 2.7 gha/person, indicating that we are placing a burden on the only truly renewable resource, biodiversity, of 1.5 times its productive capacity. In other words, we are overshooting Earth's capacity by as much as 50 % Thanks to biodiversity humanity enjoys a range of ecosystem services free of charge. If the ability to provide these services deteriorates, we will lose the ability to procure sufficient food and water from nature, the ability of nature to regulate the climate, and also cultural services such as the aesthetic qualities of nature. If we do not reduce this overshoot of 50 %, we may face an environmental Big Bang any time—indeed, considering the speed of environmental degradation seen today, we may already have pulled the trigger. In any case, we must reevaluate these seven environmental risks.

Each one of these seven risks is important and must be studied in depth, but if we look at every single one of them as an isolated environmental problem and search for individual solutions, it is clear from history that we may very well end up merely creating trade-offs. Why, for example, has the depletion of energy and resources accelerated? The reason is that the endless expansion of human activity driven by underground resource-based technologies invited the depletion of energy and resources, as well as the deterioration of biodiversity, problems with the distribution of water and food, a rapid increase in population, and climate change or global warming. Thinking of energy, we know that a human being, essentially, requires about 2,000 kcal of energy per day to exist. This figure has risen gradually as we have pursued comfort and convenience. As long as our daily lives were embedded in the cycles of nature, this was a manageable issue. But, as we discovered the magic powers of underground resources and energy, consumption started to increase exponentially. The Japanese per capita energy consumption, which around 1960 was approximately 28,000 kcal, due to high economic growth rates grew to some 92,000 kcal in 1975—a time when Japan had joined the club of industrial nations—and in 2009 had risen to 114,000 kcal. (In the United States the figure is approximately 240,000 kcal/person). What was not a risk 100 years ago, or even 50 years ago, has undoubtedly been made into a risk by the expansion of human activity. That is, we can say that the very expansion of human activity *is* the global environmental problem.

What is at stake today is how to cease or reduce this bloating of human activity. Of course, if the sole purpose were to do so, one choice might be to ration everything we consume. In a worst case scenario, we might have to choose this option, but it would entirely annul the wisdom humanity has accrued so far, as well as the educational efforts behind it. The real challenge we are facing today is how to mobilize all our knowledge and wisdom in order to cease or reduce the excessive expansion of human activity while at the same time enabling the most fundamental of human desires, to live wholesome, fulfilling lives. Furthermore, we must identify the technologies required to achieve this and create a new approach to manufacturing.

3.4   The Expansion of Human Activity

It is of course possible to define what is meant by "human activity" from the viewpoint of both social science, the humanities, and natural science, but in any case, we are talking about individual activity as well as group activity, and the social structure to which these relate. The point of departure for contemplating the nature of human activity is, surely, "the form or mould of daily life." What we have to discover today are new possible forms of living (Here, we define the "form of living" as the individual's basic philosophy of living, whereas "lifestyle" is seen as one constituting element of this).

The present economic system presupposes an endless supply of natural resources and ignores the problem of waste, and it is clearly within this framework that a rapidly increasing population combined with an endless pursuit of convenience have led to the expansion, or bloating, of human activity. There is a constant over-production of goods to satisfy our material desires, and once this has become the natural state of affairs, no one any longer notices the abnormality of it. We use money and other means to purchase food and water, housing, refrigerators, tv-sets, cellular phones, automobiles, clothes and accessories, travel services, beauty treatment and any other number of goods and services. It is this endless growth in goods and services that, in turn, translates into environmental impact.

What direction can we, then, take in trying to reshape forms of living? What does it really mean when we say that eco-products have become excuses for further consumption? Somewhere along the road, we have created a distorted understanding of the nature of prosperity. When we should really be thinking of "the level of prosperity required for people to live decent lives", we have, for some reason, instead come to believe that "people cannot live decently without a materially rich everyday life".

This, we believe, is the misunderstanding of the nature of prosperity which causes much of our present malaise (see work by Hirobumi Uzawa and Katsuto Uchibashi).

"The level of prosperity required for people to live decent lives" is a definition that allows us to start thinking about the relations between ourselves and other people, between people and nature, and, ultimately, the links between all forms of life on Earth.

From this, we can also start uncovering mindful or spirited lifestyles which focus not primarily on consumption but on achieving harmony with your inner self— gaining a sense of sufficiency and valuing the preciousness of all things. Products that encourage a spiritual fulfillment, or encourage people to enjoy entertainment in everyday life, may be more expensive, but tend to last longer. These are products that people feel affection for and want to repair over and over again; perhaps, we could say, things with a sense of time and history subtly built into them. Of course, safety is a must, but in many cases a certain degree of inconvenience may, paradoxically, help close the gap between technology/products and their user, leading to a deeper sense of affection and trust. Such products may also lead to more positive appraisals of the corporate brand under which they are provided.

The launch of the Sony Walkman in January 1979, showed the world an entirely new lifestyle in which people could take music with them outdoors. The Walkman was, we believe, a product marketed as an enabler of this new lifestyle. Is that not the very reason why the Walkman continued to be a successful product in the market over 33 years and helped make the Sony brand powerful? This can be seen as a case in which a technology actually assumed responsibility for a lifestyle.

The idea that "people cannot live decently without a materially rich everyday life", however, is the kind of thinking that has fuelled material desires. Endlessly, new products are introduced to the market, and the argument that "since it is an eco-product, buying more is better" is used to encourage increased consumption. Who, in the last instance, takes responsibility for the acceleration of environmental degradation that this leads to? Have not consumers started taking notice, though, that something has gone astray? This kind of relationship between producers and consumers, is based on a set of criteria giving preference to producers. Producers and consumers relate only through a weak link called money, and neither the face of the engineers nor the company behind the product is visible. Technology (in products) becomes a black box, and half the operation manual—often as thick as phone directories of the good old days—is filled with warning signs and comments such as "if you touch this or that, we do not take responsibility for what might happen." With this approach, even minor issues give reason for the consumer to place claims with the company, and as a result, the corporate cost of claims management rises exponentially.

Why do we manufacture products in the first place? The reason, of course, is to help create prosperous lives. But, the moment material richness itself becomes a precondition for living decent lives, we are encouraging unlimited material desire. Thus, an endless cycle of technological development and the marketing of products has taken place, merely to satisfy the unlimited material desires of citizens. The feedback system by which we have attempted to return this relationship

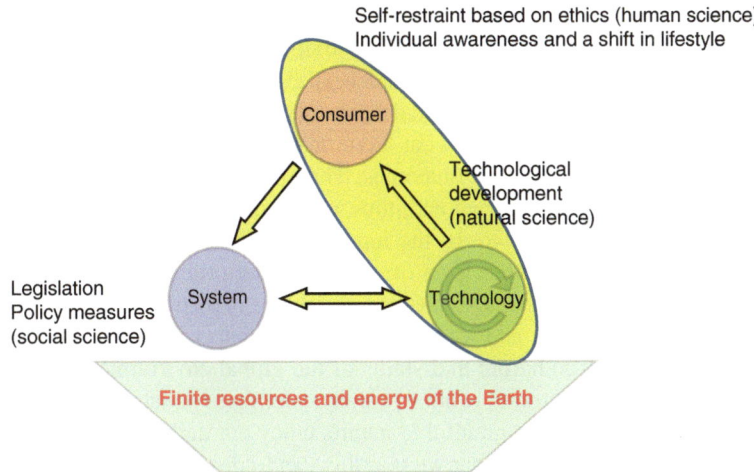

Self-restraint based on ethics (human science)
Individual awareness and a shift in lifestyle

Consumer

Technological
development
(natural science)

Legislation
Policy measures
(social science)

System

Technology

Finite resources and energy of the Earth

Fig. 3.2 In a resource constrained world, the role of technology changes significantly

between technology and citizens to a sound footing, has been policy guidance in the form of laws and regulations. So far, these three elements—technology, citizens and policy—have, apparently, interacted favorably supporting continued economic growth. Needless to say, though, this relationship can only function on the basis of an unlimited supply of the Earth's resources and of energy. Today, we have learned that this foundation for economic growth has its limits, and that we have little time left to change course. If the foundation changes, of course the relationship between technology, citizens and systems built thereon must change as well. The root cause of our problems today, however, is that this relationship has not shifted at all (Fig. 3.2). Technology, now under the banner of ecology, continues to be mass-produced, society encourages consumption through so-called eco-points or "drive ad libitum on the highway during weekends for 1,000 yen" type of initiatives, and citizens, encouraged by such convenient excuses, continue to buy more (eco)products.

It is about time that technology (corporations) start marketing products (technologies) that assume responsibility for the kind of lifestyle they help create.

The Earth uses the smallest possible amount of energy to drive a perfectly cyclical system, and through a continuous process of natural selection has created a sustainable natural world. We humans also, supposedly, existed only as one species within this environment. However, starting with the industrial revolution in Great Britain, we have used technologies based on underground resources to enable mass-production, thus creating the global environmental problem.

The unlimited material desire of humans and the technologies based on underground resources applied to meet this desire have caused the excessive expansion of human activity, and led to the seven environmental risks mentioned above. Obviously, underground resources are not, on any time scale meaningful to humans, replenished. Meanwhile, nature has a great ability to adjust and heal, and through most of history has been able to restore itself from the impact of our exploitation or waste. Recently the pace and volume of our activities, however, are such that we are exceeding nature's capacity to adjust and heal. The biggest problem is that the excessive expansion of human activity, despite much effort to the contrary, actually continues to accelerate, and thus, what were once risks will turn into threats and cause undesired phenomena and outcomes. The enormous cyclical system of the Earth is completely interconnected and, thus, environmental problems grow bigger in a chain reaction. In other words, the input that is the expansion of humanity's activity, causes an output in the form of environmental problems. As a result, the reproductive capacity of the Earth has decreased, the depletion of non-renewable energy sources accelerated, and we are traveling at full speed toward civilizational collapse. This is the true nature and status of the global environmental problem. Of course, the aim for us today is not to suffer under these constraints as we try to find a solution. We should be grateful to nature, enjoy our daily lives, and mobilize all the knowledge and wisdom that we possibly can to discover how we may cease or reduce the excessive expansion of human activity.

Bibliography

Ishida H (2009) Channeling the forces of nature. Tohoku University Press, Tokyo
Ishida H, Furukawa R, Dentsu Grand Design Laboratory (2010) Kimi ga otona ni naru koroni. Kankyou mo hito mo yutaka ni suru kurashi no katachi (By the time you grow up – a way of living that makes both the environment and people rich). Nikkan Kougyou Shimbun, Tokyo
Worldwide Fund for Nature (2012) Living Planet Report 2012. Worldwide Fund for Nature, Japan
Stern N (2006) Review on the economics of climate change. Office of Climate Change, Japan
Sugiyama M (2010) Jio-enjiniaringu gaisetsu (Overview of geo-engineering). Central Research Institute of Electric Power Industry, Tokyo
Sugiyama M (2011) Kikou kougaku nyuumon (Introduction to climate engineering). Nikkan Kougyou Shimbunsha, Tokyo
Uzawa H, Uchibashi K (2009) Hajimatte iru mirai "The future that has already begun). Iwanami Shoten Publishers, Tokyo
International Energy Agency (2010) World Energy Outlook 2010. International Energy Agency, Paris
Yamamoto R (eds) (2006) Think the Earth project. Kikou hendou +2°C (Climate Change +2°C). Diamond Publishing, Tokyo

Chapter 4
A New Way of Manufacturing and Living

Abstract In order to create a sustainable society, we must acknowledge the need for manufacturing and living that take both the Earth and people into consideration at the same time. Respect for the Earth means the establishment of a resource-cyclical society, respect for people means acknowledging their desires in life. Human desire is characterized by the irreversibility of (perceptions of) quality of life. The nature of our desires is such that once we have gained a certain level of comfort or convenience, we are not readily willing to let go of it. In order to think affirmatively of both a resource-cyclical society and the irreversibility of quality of life, we need technologies that can help bring about new ways of living. This also necessitates the creation of a new set of values, which can effectively counter the common sense of conventional technology, namely that the aim of technological development is to gain convenience and comfort with the smallest possible input of effort and time. We need an approach to technology which does not use scarce resources mined from the Earth's crust, processed at high temperature and high pressure to realize comfort and convenience while fuelling material greed, but which learns from nature using resources found in abundance above the Earth's surface, processing these at room temperature and normal pressure to provide exciting new ways of living that enable spiritual fulfillment. Nature is the only perfectly sustainable system found on Earth. Only a technology that learns from nature and has nature built into it will be able to realize ways of living that allow for truly wholesome, fulfilling lives.

Keywords Culture and civilization • Material desire and spiritual desire • Sustainable society • The irreversibility of (perceptions of) quality of life • Ways of manufacturing and living

When we talk of new forms of manufacturing and living to which humanity must now shift, what kind of "forms" are we actually talking about? Obviously, since the requirement is to achieve truly prosperous, wholesome lives while at the same time

E.H. Ishida and R. Furukawa, *Nature Technology: Creating a Fresh Approach to Technology and Lifestyle*, DOI 10.1007/978-4-431-54613-9_4, © Springer Japan 2013

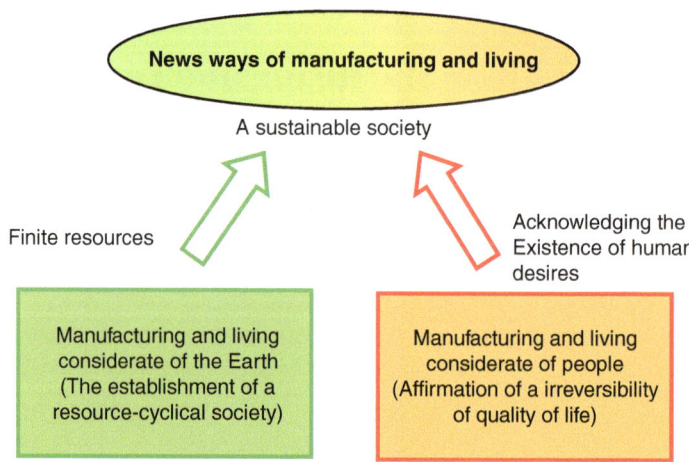

Fig. 4.1   The two elements needed to create a sustainable society

ceasing or reducing the expansion of human activity, we are talking about "a sustainable way or form of living which takes both people and the Earth into consideration". Sustainable development has been defined as "development that meets the needs of the present without compromising the ability of future generations to meet their own needs". (United Nations 1987), but in more concrete terms, there are two constituting elements when we look at "new ways of manufacturing and living"; one is making things and living with consideration for the environment (the creation of a resource-cyclical society), the other is making things and living with consideration for people (the acknowledgement of people's desires).

We can only open the door to a truly sustainable society by thinking affirmatively of both these elements (Fig. 4.1).

4.1   Manufacturing and Living with Consideration for People

What does it mean when we say manufacturing and living with consideration for people? The most important thing, here, is how to look at the issue of human desire. As mentioned above, the elements needed to create a sustainable society are both consideration for the Earth (a resource-cyclical society) and respect for the fact that people have desires. If the direction of these desires merely fuel material greed, we will clearly end up with a situation that contradicts the creation of a resource-cyclical society. Material greed is the urge to buy more accessories, more bags, a larger car, etc. Some people are satisfied by the purchase of a multitude of things and gadgets; this is the essence of material desire. In comparison, desire fulfilled by, for example, experiencing a sense of achievement in your work, or from repairing your old, oily car over and over again is a spiritual—or immaterial—desire.

Desires that can only be fulfilled by things and goods require large volumes of resources and energy and inevitably impact greatly on the environment. Can we, then, rid ourselves of our desire? Perhaps, if we find ourselves entirely unable to solve the global environmental problem, we may be forced to rationalize all that we consume, but this, obviously, is an approach far removed from our goal of achieving wholesome, truly prosperous lives. Or, maybe in an ideal world, we should all become monks or priests, forsake things and embark on a journey of spiritual hardship...These are, however, virtually impossible solutions. In Buddhism, for example, even conduct that truly benefits yourself is seen as a challenge for the individual, whereas the most important, altruistic conduct is regarded to be a highly difficult, final goal to attain. To hope for all people to be examples of altruistic conduct seems extremely unrealistic. In such a world, corporations would no longer have any role to play, and their very raison d'etre would vanish. If corporations want to avoid such a situation, there is a need to change course drastically, today.

Is it, then, possible to envision the creation of a resource-cyclical society while acknowledging our desires? To do so, we must understand the nature of human desire. Our desires display a trait that we call "the irreversibility of (perceptions of) quality of life (for details, see "Channeling the Forces of Nature" by this book's authors). We do not readily let go of comfort or convenience once we have gained it—that is the nature of human desire. We are unable to part with our cellular phone, a device we did not have 20 years ago, and if someone were to remove it forcibly, we would feel distressed. The nature of our desire is such that although we know that the phone uses numerous metals under threat of depletion, we cannot let go of it. Even if we may fancy the idea that the Edo Period (1603–1868) was an example of a sustainable society, we are unable to return to the lifestyle of that time. If we were able to return, or to let go, the global environmental problem would not have arisen. Needless to say, we are not arguing that this irreversibility of quality of life is in itself an evil; once you have experienced great music, you want to hear more of it, once you have stood face to face with beautiful paintings, you want to see more... This is the way in which culture has progressed. If, however, we can say that it is the single-minded pursuit of the material aspect of quality of life that is the root cause of the environmental problem, then we obviously must avoid fuelling this kind of desire. With desires that fuel material greed, we cannot possibly create new ways of living that take both people and the Earth into consideration. While acknowledging the existence of human desire, it is clear that if we want to look affirmatively at the creation of a resource-cyclical society, we must find a way to shift desires away from material greed towards the encouragement of spiritual desires.

If we think, as an example, of watches, this means coming across and treasuring a particular watch for life, rather than having the drawer full of cheap disposable watches. To move in this direction, each one of us needs to accumulate knowledge and gradually change our way of living. A watch that you treasure like this will probably be with you for life, and every overhaul or new scratch it gets actually deepens your affection for it. Items that you want to have with you at all times, that make you feel happy and relaxed, and that become part of you the more you use them—are those not things encouraging spiritual fulfillment? Or, rather than living in an apartment with a perfectly controlled indoor climate, maybe you would feel richer at heart—even if it is a little hot or cold at times—if your dwelling is in nature, where you can feel the wind, listen to nature's sounds, enjoy the rustling of leaves or the chirping of insects and the smell of the forest nearby? Spiritual desires can be fulfilled even at very low environmental impact, and if technologies able to meet such demands were to become plentiful in society, we would become able to satisfy people's desires with little impact on nature. In order to achieve this, we cannot take technology as our point of departure; we must start by considering lifestyles.

In any case, there are clearly a multitude of technologies recently playing to people's material greed. For example, even without significant functional changes, manufacturers of cellular phones frequently change design and encourage people to buy a new device. In order to increase sales, corporations fan people's lust to consume. As a result of this, what happens? According to our own surveys, about 40 % of the Japanese feel some kind of stress related to things. When people buy an item, they think they truly want it, but once they find themselves surrounded by things, it becomes stressful. Fanning material desires and flooding people with useless things is clearly an abnormal situation. Adding up many "small wants", people start experiencing "big stress", and ultimately their ownership of too many things mentally immobilizes them.

What does a way of living that encourages spiritual desires then actually look like? Whether this way of living is indeed possible is an issue we will leave for later, but the Japanese have both the ability to create such a lifestyle and a long history full of examples from real life.

4.2 Manufacturing and Living with Consideration for the Earth

Manufacturing and living with consideration for the Earth means creating a resource-cyclical society. In 1992, at the Earth Summit in Rio de Janeiro, the world's developed nations promised to create a sustainable society. A sustainable society— and the important factor in its creation, a resource cyclical society—are, however, despite much effort further from realization than ever before. Let us take a look at a resource-cyclical society from the viewpoint of technology (Fig. 4.2). Here, we must realize that we have become unable to live in symbiosis with nature.

About 10,000 years ago, when humanity began living in permanent dwellings, we started erecting a wall—also called a system boundary—between the natural ecosystem and the human ecosystem in which we live. As civilization developed, this wall became ever higher and sturdier and humanity ever more removed from nature. If we could live entirely within the system boundaries of nature, there would not be any environmental problem, but unfortunately, we have become unable to do so. To maintain the human ecosystem, we import resources and energy across the system boundary (this is called input), reshape and use these to gain comfort and convenience, and, then, when we can no longer find use for them, toss the remains

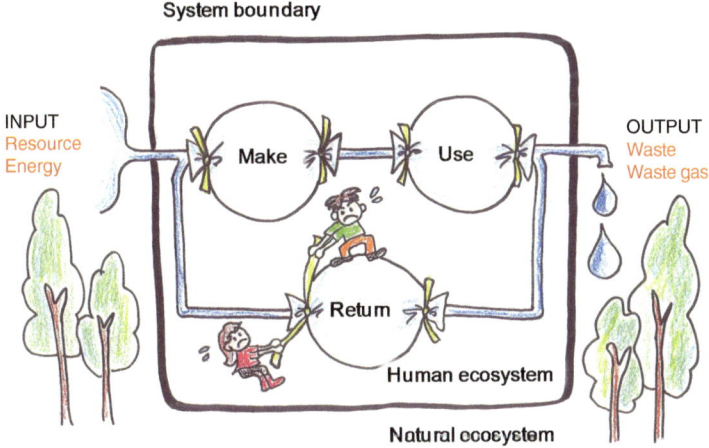

Fig. 4.2 Reducing resource inputs is the foundation for a resource-cyclical society

back into the natural ecosystem as waste (this is called output). Such a relationship has continued for a very long time. As a result thereof, some of the things we threw away in nature have come the whole cycle round and are now reentering from the input side giving birth to the pollution problem. Today, we have to create a resource-cyclical society that does not burden the environment—the question is, how? This, of course, is achieved by decreasing the volume of output as much as possible, but the essential thing here is that in order to do so, we must first decrease input to the largest possible degree. When input is smaller, output will be so too, and then the important task becomes how to return the much reduced volume of output to the resource-cycle as new input. As long as we merely make efforts to reduce output, without dealing with the task of reducing input, this loop will become constipated and, in the end, burst.

4.3 Ways of Living (Lifestyles) Incorporating Nature

In order to take a positive approach to both a resource-cyclical society and the irreversibility of quality of life, we need technologies that can reform and reshape ways of living. If we can say that modern technology has aimed to achieve its purpose with the smallest possible effort and input of time, and without using artistic skills and techniques, then what is needed today is a new type of technology, which based on a fresh set of values can negate this approach to problem solving in life.

Let us once more contemplate the nature of conventional, modern technology. The main technologies supporting modern society are underground resource-based technologies born after the industrial revolution. Apparently, we are unable to utilize these technologies, which did not exist on Earth before the industrial revolution, in a truly effective manner. As already mentioned, the history of technology is a history of trade-offs. Freon gas, by some praised as the greatest invention of the twentieth century, in fact destroys the ozone layer and is a greenhouse gas with an astonishing 20,000 times the global warming potential of CO_2. The generation of electricity so vital to our lives, and convenient means of transportation such as automobiles and airplanes all add to global warming...Technologies that were developed in good faith, turn out to cause terrible problems. It is the continuation of this process that has led to today's serious environmental problems. Of course, we cannot entirely disregard the value of underground resource-based technologies. Is it not rather the time, now, to contemplate which ones of these technologies we truly cannot do without? Now is also the time to consider in earnest what new types of technology we need that do not rely on high temperature and high pressure in highly complex processes, as is the case with underground resource technologies.

Looking back at the history, our species, *Homo sapiens sapiens*, was born some 200,000 years ago and has lived most of the time since then as hunter-gatherers. About 10,000 years ago, humanity mastered farming (the agricultural revolution), and some 200 years ago spurred on by the industrial revolution in Great Britain, entered the industrial age. If we think of these 200,000 years as 1 month, the agricultural revolution took place 1.5 days ago, and the industrial revolution a mere 43 min ago. Our excessive practice of agriculture has transformed nature, and in as little as 200 years industry has ravaged energy and resources created by the Earth over a period of hundreds of millions of years or more. To gain but an instant's worth of pleasure we have pursued material wealth, while disregarding future generations and ignoring what legacy we might be leaving behind for them. Stupefied by the almost narcotic effect of material wealth, we yearn for even more pleasure, and, increasingly, it looks as if we have fallen into a hell of endless, expansionist behavior. You cannot help but thinking that maybe, from the viewpoint of nature, humans are a species that should have been weeded out by natural selection. The only way to avoid this happening, is for us consider how to reintroduce our civilization into the cycles of nature.

As a result of the estrangement from nature caused by the industrial revolution, and driven by our use of underground resources, our pursuit of convenience is no longer based on our own efforts, but has been "outsourced" to the tools of technology. Needless to say, the resource efficiency of this outsourcing is worse than if each of us were to make our own efforts and has led to an exponential growth in the consumption of resources and energy. Take, for example, Dell, who years ago started outsourcing part of its manufacturing to the Taiwan computer manufacturer Acer. Happy to find that short term profits went up, Dell continued to expand this outsourcing until, one day, the company had lost all its technological ability to manufacture. This almost tragicomical situation is similar to what humanity is doing on a global scale and at full speed. Using money as a tool, we outsource all and everything, and today appear to believe that we can even buy time with money. Frozen and instant foods make us forget what food in season tastes like; air conditioned highrise buildings with windows that don't open make us forget the seasons or the smell of the wind, as well as the sounds of chirping birds. Black box technology rapidly

Fig. 4.3 Average life span of Japanese (men) and energy consumption. Created on the basis of "Survey for Vital Statics" (Ministry of Health, Labor and Welfare 2012) and "Energy white paper" (Agency for Natural Resources and Energy 2012)

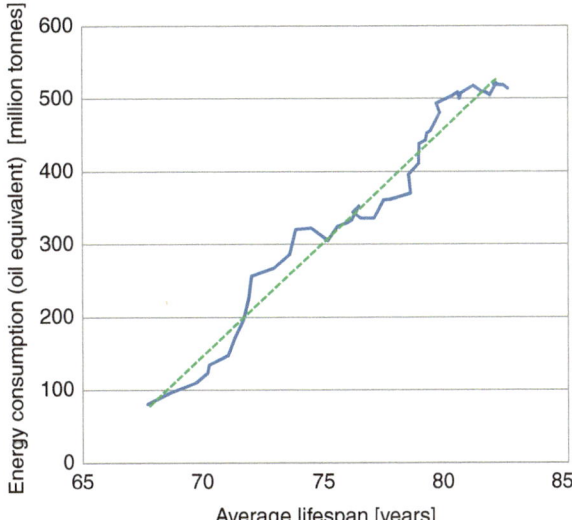

causes us to forget—or unlearn—the wisdom and skills we once possessed. What we have to consider today is how, little by little, to regain some of what we have lost through outsourcing.

The only way for humanity to survive is to find ways to reintroduce ourselves into the cycles of nature. This challenge depends on to what degree we can reduce outsourcing, and on whether we can find other measurements in life than money.

If you look at the relationship between weight and the life expectancy of mammals, a human being weighing 60 kg should, it is said, live for an average of some 30 years. Our very existence, obviously, deviates from the rules of nature, but if you consider the fact that around the end of the nineteenth to the beginning to the twentieth century average life span was about 30 years, can you maybe say that humanity until then lived within the great cycles of nature (Fig. 4.3)? Whether or not this is true, our life expectancy has grown rapidly in the last century—the average for Japanese men exceeded 60 years in 1951, 70 in 1971, 75 in 1986, and in 2010 had reached 79.6 years. This increase in life span is a welcome development, but in the same period energy consumption has increased by a factor of 3.6, and records show a very accurate correlation with life expectancy (see Fig. 4.3). If we contemplate the fact that longer life spans are intricately linked to higher energy consumption, it becomes increasingly clear that we are expanding the length of our lives while gobbling up vast volumes of resources and energy.

At least, it's fair to say that if humanity continues along its present path, the future of the human species on this Earth looks precarious. We have to stop now and ask ourselves whether we are really prosperous today. The answer, obviously, is no. Even just continuing society as it is today is becoming difficult. In our belief of continuous progress we have made great efforts, but rather than actually achieving progress, we have driven ourselves into dangerous circumstances. An additional issue that we must consider is the fact that the present problems were created by the 20 % of world population belonging to overwhelmingly rich nations, drawing with us down the remaining 80 %, who do not even share the benefits of modern technology.

The Great East Japan Earthquake which occurred at 14:46 on March 11, 2011, taught us we have forgotten that humans exist as part of nature's great cycles, and also that we have forgotten it is nature which keeps us alive. The earthquake awoke us from our illusion of being able to conquer nature. We could not even extinguish the "fire" at the Fukushima Dai-Ichi Nuclear Power Plant; a state-of-the-art seawall erected over 30 years at the cost of 120 billion yen, and in 2010 included in the Guinness Book of World Records, was destroyed by the tsunami; and many houses and infrastructural systems were utterly devastated in an instant. Having to live without water, electricity or gas, showed us, with the force of fear as teacher, an entirely different world from the one we knew just the day before. Returning the energy supply to the level experienced before this disaster will probably be of the utmost difficulty. In this situation, a paradigm shift is being called for that would enable us to avoid returning to the previous state, and to realize new lifestyles and technologies allowing us to live wholesome, fulfilling lives with less energy and fewer resources.

The Earth was born around 4.6 billion years ago and is expected to continue its rotation for another 5 billion years or so. Life was born some 3.8 billion years ago, and since then has continued a process of natural selection to create a perfectly cyclical and sustainable system driven with the smallest possible input of energy. We ourselves are on the verge of pulling the trigger of a civilizational collapse, but even if civilization were to fall and huge amounts of species to be lost due to our conduct, nature will most likely continue to maintain a sustainable system. That is how diverse and perfect a system nature is. If we continue as today, humans, on the other hand, will lose the renewable ecosystem services provided by nature as well as non-renewable mineral and energy resources, moving straight towards civilizational collapse. The crucial thing is that we will become unable to enjoy the blessings of nature upon which our very existence depends.

It is said that wisdom derives from culture, while it is the accumulation of technology that has created civilization. Normally, culture and civilization should develop and grow as one entity with each element stimulating the other. However, since our technologies started becoming black boxes, and we, the users, have become unable to understand how they function, civilization has started leaving culture behind and is developing in a reckless manner. Why did this happen? The very reason, we believe, may be the absence of a proper view of nature. The technologies supporting our underground resource-based civilization are products of the era after the industrial revolution. It was the notion of man ruling nature, and of man's parting with nature, that allowed such technologies to flourish. In the mechanical view of nature espoused by Francis Bacon and Rene Descartes, nature is expressed in mathematical formulae. The technological paradigm built on the notion that if only humans can master these formulae we may exploit nature as a slave, has ruined nature in order to satisfy our desires and invited the present crises of resources, energy, biodiversity and abrupt climate change.

It is high time for us to act to shift from this underground resource-based civilization to a civilization of life with new ways of living and manufacturing incorporating a sound view of nature. In order to do so, we must learn from the only truly sustainable system on Earth, nature, to create technologies with a view of nature as their foundation. Doing so, we will give birth to new forms of living that are not merely pursuing convenience, but where the faces of truly happy and fulfilled citizens using technology to create new prosperity will be visible.

A first step in this process is to clarify what wholesome, fulfilling lifestyles might look like under the severe environmental constraints we are facing, and then consider what technologies are required to realize these.

Bibliography

Agency of Natural Resources and Energy (2012) Energy Whitepaper, Agency of Natural Resources and Energy, Japan

Ferguson ES (1994) Engineering and the Mind's Eye, MIT, reprinted version, Boston

Hiroi Y (2003) Seimei no seijigaku (The political science of life). Iwanami Shoten, Tokyo

Ishida H (2009) Channeling the forces of nature. Tohoku University Publishing, Sendai

Ishida H, Hideto N (2010) Enjinia no tame no kougaku gairon, daikyuu kou (Introduction to engineering for engineers, lesson nine). Minerva Shobo, Tokyo

Ishida H (2010) Atarashii kurashi to tekunorojii wo kangaeru iinkai (Committee on NewWays of Living and Technology). In: Chikyuu ga oshieru kiseki no gijutsu (Miraculous technology the Earth teaches us). Shoudensha, Tokyo

Kitou H (2000) Jinkou kara yomu nihon no rekishi (Japanese history from a demographic viewpoint). Kodansha Gakujutsubunko, Tokyo

Meadows DH, Meadows DL, Randers J, Behrens WW (1972) Limits to growth: a report for the club of Rome's project on the predicament of mankind. A Potomac Associates Book, Vermont

Meadows DH, Meadows DL, Randers J (1992) Beyond the limits. Chelsea Green Publishing, Vermont

Ministry of Health, Labor and Welfare (2012) Survey for Vital Statics, Ministry of Health Labor and Welfare, Japan

Nakaoka T, Suzuki J, Tutumi I, Miyachi M (2001) Sangyougijutsushi (A history of industry). Yamakawa Shuppansha, Tokyo

Nielsen KS (1984) Scaling: why is animal size so important? Cambridge University Press, Cambridge

Ohgushi T (ed) (2003) Seibutsutayouseikagaku no susume – seitaigaku kara no apuroochi (For the advancement of a science of biodiversity – an approach from the field of ecology). Maruzen, Tokyo

Ohnuki T, Sakashita K, Seguchi M (2002) Kougaku rinri no jouken (The criteria of an ethics of engineering). Koyo Shobo, Kyoto

Papanek V (2005) Design for the real world: human ecology and social change. 2nd revised edn. Academy Chicago Publishers, Chicago

Schmidt-Bleek F (1997) Fakutaa ten – ekokouritsukakumei wo jitsugen suru (Factor 10 – realizing the eco-efficiency revolution), Trans. Ken Sasaki. Springer, Tokyo

Tomonaga S (1979) Butsurigaku toha nan darouka (What is physics? (vol. 2)), Iwanami Shinsho, Tokyo

United Nations (1987) Report of the World Commission on Environment and Development: Our Common FutureOur (Brundtland Report)

Van Andel TH (1985) New views of an old planet: continental drift and the history of the earth, 1st edn. Cambridge University Press, Cambridge

Yamamoto R (ed) (2007) Think the earth project, Ikimonogatari (Stories of living beings). Diamond Publishing, Tokyo

Chapter 5
Forecasting and Backcasting

Abstract Will we be able to create new lifestyles extrapolating from the present? If we look at so-called lifestyle hazard maps, this looks difficult. With thinking based on forecasting, which is our conventional approach today, we will not find is easy to change lifestyles. Trying to envision how, based on our present way of living, we might become able to balance consideration for the environment with our desire for a wholesome, fulfilling life is a task that is very hard to solve. As a result of this, we would only be able to find partial solutions and would end up creating eco-dilemmas. What is needed today is thinking based on backcasting. This is thinking which focuses on optimization of the total system, taking the severe environmental constraints likely in 2030 as a basis for the contemplation of wholesome, fulfilling lifestyles.

In order to enable lifestyle design based on backcasting, we must clarify the environmental constraints of 2030, envision what society might look like under these constraints and, based on this, outline what wholesome lifestyles might look like. Using this methodology, the authors of this book have so far outlined more than 1,500 different lifestyles, and using these we are able to uncover the constituting elements of wholesome, fulfilling lifestyles.

Keywords Backcasting • Constituting elements of lifestyles • Environmental constraints • Forecasting • Lifestyle design method • Lifestyle hazard maps • Lifestyles

5.1 Lifestyles Based on Forecasting

It is not easy to change lifestyle. It is quite natural that if people, for example, are asked to change from commuting to work by car to the more inconvenient public transport, they will not want to make the shift to greater inconvenience. People who have used the tv-set or air conditioner without paying much attention to energy use will, most likely, find it difficult suddenly to change to a lifestyle in which they

E.H. Ishida and R. Furukawa, *Nature Technology: Creating a Fresh Approach to Technology and Lifestyle*, DOI 10.1007/978-4-431-54613-9_5, © Springer Japan 2013

conserve energy and refrain from using the tv-set and air conditioner. There are many such hurdles when we try changing lifestyles to realize wholesome, fulfilling living with low environmental impact.

One such hurdle is the way in which people tend to desire a lifestyle that is very much an extension of the present. Let us call this "thinking based on forecasting." People own and build their everyday lives on consumer goods such as air conditioners, refrigerators, tv-sets, lighting, as well as microwave ovens and personal computers. In a majority of cases, it is expected that products already on the market will be further upgraded, and that the consumer then buys new appliances with enhanced convenience or more enjoyable functions. Or, consumers buy anew when the appliance they were using breaks down and needs repairing. Purchasing new appliances in this way, however, does not alter people's lifestyle. Buying anew does not in any way diverge from the present way of living based on consumer goods; on the contrary, it helps cement a lifestyle which uses things to achieve satisfaction. Thinking based on such forecasting—that is, extending in a straight line from the present—is not aiming to change this pattern of behavior. Rather, corporations encourage this kind of forecasting in consumers and may even be said to conduct business in such a way that the barrier to change lifestyle is raised. "Changing lifestyle(s)", may refer to either quantitative or qualitative changes. When we talk about changes in lifestyle, one common interpretation refers merely to reduced frequency of usage. Changing lifestyle, however, does not only involve changes in how often things are used, but also qualitative aspects of change such as "how things are used" or considerations of "what added value can be gained from using things (in new ways)."

If we lower the frequency of use of an item, environmental impact will be reduced. For example, if we simply use an air conditioner less and thus act to conserve energy, environmental impact will fall. There is another approach, though, in which we by changing the qualitative aspect can add value while also reducing environmental impact. Take, for example, the above-mentioned shift in commuting by car to using public transport. By using public transport, we may well be able to discover things we had not noticed driving a car, thus gaining more chances to watch the trees or beautiful flowers in bloom along our route. In this case, our shift in commuting is a lifestyle change that both adds new value to our lives and reduces energy consumption considerably. That is, we can at the same time increase the environmental efficiency of our lifestyle and reduce environmental impact. Needless to say, how—and whether or not—added value changes differs from person to person.

While it is indeed possible both to reduce impact and increase efficiency by changing lifestyle in qualitative ways, reality often is that we have difficulty overcoming the barriers to altering lifestyle and thus often end up merely conserving energy, reducing usage or taking other measures to reduce environmental impact. In this case, we end up pursuing a lifestyle that is only a partial solution, focusing on the environmental aspect alone.

5.2   Measuring the Impact of Forecasting

Lifestyle Hazard Maps is a new methodology which, while taking the needs of people and society as well as environmental impact into account, allows us to map lifestyles that may be unsustainable. Lifestyle Hazard Mapping was conceived as a tool for use by countries or corporations to plan technological development strategically or for the purpose of policy making. It is a strategic map which on the X-axis plots consumers' preference for different consumer goods, and on the Y-axis the lifecycle environmental impact of the same goods. Based on this research, it is possible to make a quantitative analysis of thinking derived from forecasting. Research by Masuda et al. show that even when assuming severe environmental constraints, the preference of consumers for different consumer goods changes little, and that no large reduction in environmental impact can thus be expected (Fig. 5.1).

In the research, consumer preference in a situation assuming environmental constraints was defined for participants as follows: "Assume a situation in which environmental problems get worse impacting your everyday life negatively due to environmental constraints, meaning that you will have to let go of some convenient appliances." Estimations of the potential reduction in environmental impact from lower frequency of usage showed that, when adding up figures from both durable and non-durable consumer goods, a 31.2 % reduction from present levels in environmental impact was achievable. What this means is that a partial solution focusing only on the environmental aspect appears to enable a reduction of 31.2 % In an approach like this, however, where people have to endure less usage, this is about

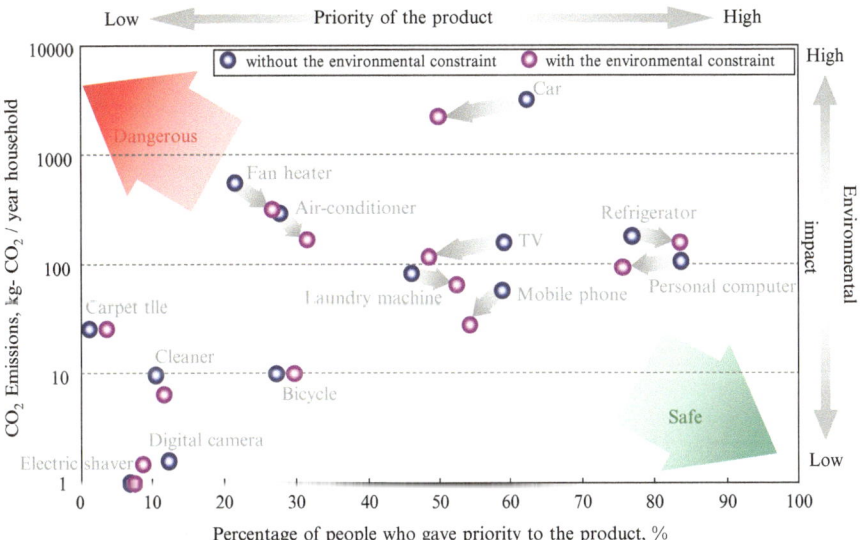

Fig. 5.1 Lifestyle Hazard Map for durable consumer goods

the limit of possible reduction. Large scale reductions of 50–80 % CO_2 will be almost impossible to achieve solely with the "endurance approach" (of lowering frequency of usage).

5.3   The Dominance of Forecasting

Progress in science and technology naturally builds on past accomplishments, but the thinking and behavior of people do not necessarily have to be dominated by forecasting, nor do we have to pursue lifestyles extrapolating from the present. Without noticing, however, we have come to be thoroughly dominated by thinking based on forecasting.

As noted above, if we lived a world with a limitless supply of resources and energy, forecasting would pose no problem whatsoever, but living under severe environmental constraints, forecasting becomes an obstacle to the potential reduction of environmental impact from lifestyle changes.

Corporations that only use a forecasting approach to do business aiming at satisfying consumers' present needs, will constantly face eco-dilemmas. The reason is that even if a company provides products or services with a supposedly low environmental impact, these are merely used to meet the ever expanding dreams and desires of consumers, without trying to limit this insatiable expansion. As dreams and desires keep growing, and under the influence of an increasing world population, new products with more functions take over from present ones and are marketed for personal ownership—all of which leads in the direction of higher environmental impact. This is one of the eco-dilemmas confronting corporations. Because of this forecasting mentality, air conditioners, refrigerators, tv-sets etc. are all developed to incorporate more functions and sold for private ownership. In a relatively small global community, this was a possible approach, but with the emergence of new economies such as China and India and a burgeoning world population, we are about to feel the true impact of such eco-dilemmas. Global environmental problems have their root cause in this expansion of human activity.

In order to create a sustainable society, a highly important task for us is to free ourselves from the rule of forecasting, shift in a direction that does not merely extrapolate from the present, and consider how to realize wholesome, truly prosperous lifestyles.

5.4   Lifestyle Design Based on the Backcasting Methodology

Backcasting is a methodology which can help us break the domination of forecasting. The backcasting methodology was first proposed by Amory Lovins in the 1970s as a way of planning electricity demand, and—under the names of "Backward-looking analysis (Lovins 1976, 1977) or "Energy backcasting" (Robinson 1982)—was used primarily in the field of energy policy. John B. Robinson writes that the purpose of backcasting is not to create a blueprint, but to show several different possible energy futures, including an interplay between social, environmental and political factors, as a basis for future goal setting or policy-making (Robinson 1990). In this way, backcasting was originally mainly used as a policy analysis tool for the exploration and evaluation of the future of energy supply.

Later, backcasting was applied as a conceptual, methodological tool focusing not only on how to reach a desired future state, but also on how to avoid or deal with possible undesired futures. Later, the methodology has also been used in the sustainability field to contemplate sustainable transport systems (Hojer and Mattsson 2000) or corporate sustainability (Holmberg 1998). With this methodology starting to catch on, Dreborg in "Essence of Backcasting" describes the usefulness of backcasting and compares it with forecasting, stating that while forecasting is based on dominant trends, and therefore is unlikely to generate solutions that would break these trends, backcasting—being a normative approach to problem-solving—is more easily applicable to long term problems, including long term sustainability issues, and is better at projecting a future image useful in creating consensus or making decisions for policy makers and the general public. He further describes how backcasting is a favourable approach when the problem studied is complex in nature, when there is a need for major change, when dominant social trends are part of the problem, when the problem of externalities is not properly dealt with by the market, and when the time horizon is long. (Dreborg 1996).

Backcasting in the 1990s started being applied in a wider setting as a methodology for the participatory envisioning of possible futures. In 1993, Vergragt and Jansen write that, as a point of departure, the backcasting methodology creates a solid image of the future state of society after which the measures needed to reach that future are considered. This future image is neither a scenario as created with forecasting nor an actual product, but has to be a firm image which appears acceptable to engineers of the present (Vergragt and Jansen 1993). Amid this, projects including a sustainability aspect and with stakeholder participation, such as the Sustainable Technology Development Programme (Weaver et al. 2000) or SusHouse were conducted, and the backcasting methodology started being used in a variety of fields.

The nature of backcasting is such that, in most cases, a time frame of 25–75 years is used and by comparing such analyses with the most reliable long term forecasts, backcasting can help clarify future goals. In this way, backcasting and forecasting methodologies by some are seen as complementary approaches (Hojer and Mattsson 2000).

Some shortcomings of the backcasting approach include the fact that the user of the methodology is easily influenced by present policy (Robinson 1982); that the methodology on its own is insufficient in envisioning possible futures; that there is a need to include an action plan (Vergragt and van derWel 1998); and that it may result in an outcome of more limited scope than when using multiple scenarios to think of the future (Masui et al. 2007). Because of these concerns, the ideal approach may be, as Hojer and Mattsson (2000) pointed out, to combine the use of backcasting and forecasting in a complementary manner.

Based on the backcasting method, the authors of this book developed a "lifestyle design methodology" to enable our own original design of lifestyles (Fig. 4.2). This is used to contemplate how people might live wholesome, fulfilling lifestyles even under severe environmental constraints. This design methodology is used to look at business strategies or policy proposals that would enable the products, services, institutions and policies required to bring about the envisioned lifestyle.

A first step in this approach is to learn about future environmental constraints. The purpose here is not to study detailed projections or data, but rather to get a grasp of big trends and major changes likely. For example, an exercise may start by using reliable, national data to make a quantitative analysis of Japan's population, energy and resource demand, climate change, water and food related issues, and biodiversity in 2030, when environmental constraints are likely to become severe. This data, however, is not easily grasped by the human brain. There is a need to study major trends so thoroughly that participants really internalize the knowledge and start using it without being conscious thereof.

The second step is for participants to imagine what society would be like under these severe environmental constraints. For example, if the price of gasoline were to rise threefold, one might imagine that fewer people would use cars, that roads would be less crowded or that there would be fewer lanes for traffic. Or, if electricity prices were to surge, one might imagine that it would become uneconomic to store large volumes of food in refrigerators, and that some people would therefore stop using the refrigerator and buy dried food instead. If this were to happen, people's homes might be filled with different smells. In this way, the aim is to try imagining as concretely as possibly what society would actually look like. Using this approach, the participant uses imagination to outline the state of society as derived from the environmental constraints. This is only an exercise in thinking in a backcasting mode—more or less like warming up exercises before running a marathon. Just as you cannot run well without warming up, when you are thinking of future lifestyles you need to train your way of thinking to enable your brain to use backcasting. It is important not only to understand the trends of future environmental constraints, but also to imagine in detail what society might be like. By taking this step of imagining the future state of society, the participant finally starts feeling as if being in the year 2030. Actual ways of living under such circumstances will then most likely start to pop up, even if still vaguely, in your mind.

The third step is the actual backcasting. Starting from the imagined state of society in 2030, participants search for problems that appear likely to occur if the present trends continue. This is the essence of backcasting—to look at the present from the viewpoint of the future. As you place yourself in 2030, in order to explore solutions (ways of overcoming barriers) to the problems just discovered, you again look back at the present. You think of what solutions your "2030 you" has been able to find to the problems faced. Here, it is important to be as concrete as possible about potential solutions. Without thinking in concrete terms, you will be unable to paint possible solutions for everyday life. Ways of living differ vastly depending on whether the person in consideration is a company employee, a designer or an author of novels; a young woman or a middle-aged man; is living in a big family or alone;

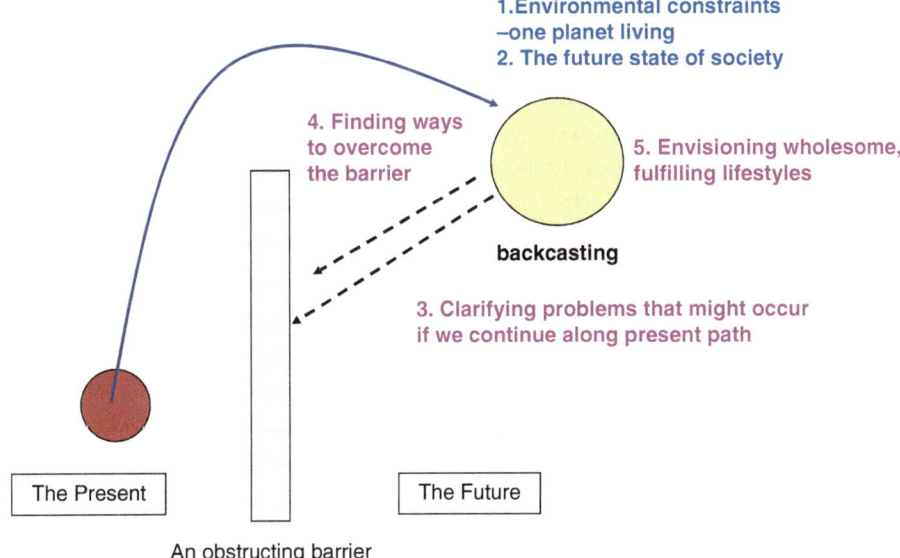

Fig. 5.2 The steps of backcasting

lives in the city or in a rural area, lives close to a station or not, owns a car or not, etc. Unless the participant thinks in as concrete terms as possible about who lives under what conditions, it will be impossible to paint a picture of possible lifestyles. From here, the participant imagines, again in concrete terms, what products, services, institutions and policies are required to enable the envisioned lifestyle.

The final step is to fine-polish the picture of a future lifestyle which takes into account the problems discovered and incorporates solutions to these. There are many different ways to portray this lifestyle—in writing, drawing or painting, or using keywords. But in any case, important constituting elements of the representation must include the imagined "future state of society", "the problems" likely to occur under future environmental constraints, "ways to overcome such hurdles or barriers (solutions)", and a vision of "wholesome, fulfilling living" and "the new values accepted by citizens." It is vital to include all of this in such a way that the attractiveness of the new lifestyle is accentuated. The presentation must be such that when other people read or see it, they are able to create their own images of "the products that might be helpful in such a way of living", or of "the policies that will be required" (Fig. 5.2).

In collaboration with universities and corporations, the authors of this book have so far outlined some 1,500 different lifestyles using this methodology. Participants have included people of many different ages, occupations and skills: university students in their early 20s, corporate engineers, salespeople, designer, creators etc. As we repeatedly conducted these lifestyle design exercises, it became increasingly clear how stubbornly the forecasting mode of thinking is ingrained in people.

Even when you have once experienced thinking in the backcasting mode, you may easily relapse into forecasting with just a moment of inattention.

Lifestyles are ways of living. The lifestyles we are trying to envision through this exercise are not mere descriptions how to live, but are at the same time solutions to the environmental problem. They are not, however, meant to be descriptions of technology or policy initiatives. Furthermore, lifestyle design, as the name suggests, includes a design element, and must be created including a high degree of novelty. Also, since we are backcasting from a specific point of time in the future, the envisioned lifestyles must also be something that could potentially be realized within the timeframe imagined. Therefore, participants are not allowed to include many highly unrealistic technologies based on pure time travelling.

Also, expressions of things that are too specifically focused on solving one isolated issue, such as, for example, "walls that absorb smells" merely point to the problem and must be avoided. It is also not allowed to be looking, single-mindedly, for novelty and new combinations. During the exercise, participants may also suddenly notice that they are thinking about issues or things that do not relate to the backcasting itself. Sometimes, participants are also caught up in focusing their attention too much on future lifestyles thus thinking that the values of people will shift more than is often the case. Painting a picture of an attractive, future lifestyle is more difficult than most people imagine. Thus many people often end up writing, for example, that a certain activity "would be enjoyable" without feeling sure whether this would really be the case. Truly enjoyable ways of living are mostly self-evident and do not need expressions such as "enjoyable" or "fun" attached to them, so there is a need to be cautious that the exercise does not end up merely playing with words. A key point is to describe the envisioned lifestyle in concrete terms. Take, for example, a lifestyle based on "local production/local consumption." This rather vague notion dominates the participants' thinking, who are often only able to come up with rather ordinary elements of the lifestyle, such as "growing your own food in kitchen gardens." The idea of kitchen gardens is still too vague. Unless you focus more specifically on the unfolding of the lifestyle, such as "who plants what where and when", it will not be possible to create truly new ways of living. Finally, participants must feel a sense of responsibility for the lifestyle they have designed— that is, it must be a lifestyle that excites them personally. Irresponsible lifestyles tend to be neither attractive nor very realistic. On the other hand, when participants are from corporations, the organization they belong to often weighs heavily on them, and most of the ideas popping up relate to the survival of this organization. As a result, these participants end up returning to the forecasting mode of thinking. Balance is what is important here. With the proper sense of balance between, for example, creativity and responsibility, imagination and reality, it becomes possible to design lifestyles that are truly wholesome and fulfilling.

Ten characteristics of lifestyle design:

1. Lifestyle design provides solutions to the environmental problem
2. Lifestyle design is not a description of technology or policies, but a vision of new ways of living
3. The aim is to create a novel approach to living

4. Technologies that are clearly unrealistic should not be included in great numbers
   One should avoid things that merely refer to specific problems
5. Looking for new combinations of things that already exist is not backcasting
6. You start by outlining your desired lifestyle based on your present values
7. Attractive wording, such as describing something as "enjoyable" or "fun" alone is not sufficient
8. The devil is in the detail—be specific
9. Be responsible for the lifestyle you design
10. Don't be burdened by consideration for the organization to which you belong

Below are some examples of lifestyles outlined using this methodology. It is also possible to create a "social acceptability index" to evaluate the extent to which people actually desire a particular envisioned lifestyle.

5.4.1 *<Living in Accordance with the Seasons>* *(Social Acceptability of Lifestyle: 76 %)*

On hot summer days, you live with the window open. Since you have a bamboo curtain which also generates solar power hanging outside your window, you are shaded from the Sun while generating energy. You regularly water the garden to make the breeze that comes through the blinds cooler. It is a yearly custom to prepare the solar bamboo blinds before summer arrives. When autumn arrives, you place the solar bamboo blinds on the roof where they will, only, be generating energy. In winter, on the other hand, you want as much sunlight as possible to enter the house, so you place a mirror in the garden to reflect sunlight up unto the ceiling of the room near the windows. The sunlight is reflected only on the ceiling just near the windows, so you are not blinded by the light while in the room, and the reflected light helps warm the room. On the ceiling near the windows, black wall paper is used. Walls inside the house are, where possible, moveable partition walls, so you can change the size of the room depending on the season. On summer days, you move the wall northerly to be able to be away from the sunlight, and in order to create a draft through the room, the walls are set at an angle to create openings. In winter, it is warmer to be near the windows, so although you may use heating, the wall is moved southerly and the living room made smaller in order to warm up less space. This is a lifestyle which changes in accordance with the seasons.

5.4.2 *<Shared Batteries> (Social Acceptability* *of Lifestyle: 70 %)*

Sometime in the past, people started moving from large family houses to apartments to free themselves from various family obligations, and gradually the nuclear family was born. Life in traditional, long family houses was abandoned, and people started treasuring privacy. In the course of these developments, the bonds that existed in community vanished.

The installation of "community-shared batteries", which allowed people to protect their privacy while sharing energy infrastructure, is one development that has started putting a brake on this trend. In this system, solar panels are installed on each housing unit in the community, and the electricity generated is accumulated in storage batteries in each building. When a family, however, travels or for other reasons uses less energy than generated, the excess energy is automatically stored in common storage batteries. These storage batteries are of help to inhabitants when rainy days continue. Excess energy is not sold to the utility company, but, based on a Japanese spirit of generosity, energy that is left over belongs to all. Thus, a community which both protects privacy and enables mutual support is born.

5.4.3 *<The Talking Refrigerator> (Social Acceptability of Lifestyle: 26 %)*

You buy a relatively small refrigerator, and when it is installed in the house give it a name. From the moment of installation, the artificial intelligence of the refrigerator is activated. When foodstuff, beverages and seasonings with barcodes are placed in the refrigerator, a sensor is activated and the refrigerator records the freshness and consume by… date. The refrigerator is like a best friend. When work finishes and you come home and touch the refrigerator, it starts talking. "You seem to be tired today. What about having pork for dinner?" Your family does not store vegetables in the refrigerator, but the sensor still registers the freshness of the vegetables placed in a box next to it. "You should eat the vegetables soon, or they will turn bad." The talking refrigerator may seem a bit impertinent at times, but since it has been set on "Edo-mode" (referring to the Edo Period, 1603–1868), you can enjoy a spirited conversation. With previous refrigerators, you most often thought of buying a new one when something broke down, but since you bought the talking refrigerator, you have become more inclined to think of having it repaired.

5.4.4 *<The Purifying Box> (Social Acceptability of Lifestyle: 28 %)*

Many people find it hard to dispose of their cellular phone or other small gadgets that they have used with affection for years, even if it is for recycling. Maybe it is because they feel a bit like amulets. Things that have been used over a long period of time are not easy to throw away. People nowadays use most things longer than in the past, but this means the house is filled with items of which the owner does not easily let go. It is here that an interesting new service, the purifying box, has come to become popular. When the time has come to dispose of the cellular phone or other small devices, more and more people use a religious purifying service to part with their beloved item in an acceptable manner (it is custom in Japan to use purifying services of temples and shrines in daily life). People who would like to clear their minds of bad memories associated with the cellular phone, bring the phone to their local temple at New Year's to get it purified along with the usual amulets and talismans. In towns that have introduced this service, the resource recycling system has started working well. The former owner can then, with good conscience and a clear mind, start looking for new things to purchase.

5.5   The Backcasting Approach Versus the Forecasting Approach

Let us once more sort out the difference between a backcasting and a forecasting approach (Fig. 5.3). The basis for conventional thinking is forecasting. That is, thinking that takes the present as a point of departure for thinking of the future.

With this kind of thinking, people try to realize both a prosperous life and the protection of the environment on the basis of their present desires (lifestyles). People who are at the peak of prosperous living may argue that "since this is wealth I earned myself, I am entitled to use it the way I want to!", while people who live at the peak of environmental concern might say that "it is a matter of course that polar bears are more important than human beings." Where, then, is a technology that is both considerate of the environment and enables a wholesome, fulfilling lifestyle positioned? We can only find it in the area where the two peaks intersect (Fig. 5.3, left). The result of this approach will be partial solutions and the creation of eco-dilemmas.

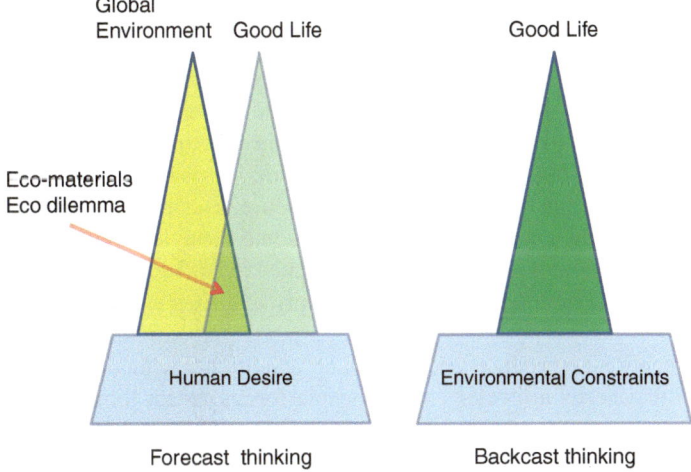

Fig. 5.3   The backcasting approach versus the forecasting approach

What, then, if one were to take a backcasting approach to the same issue? This would mean contemplating prosperous living with the severe environmental constraints of around 2030 as a foundation (Fig. 5.3, right). In this case, there is only one mountain, and the solution, of course, will be a total system approach. We like to think of these constraints as "a beneficial hoop" (holding society or community together); this hoop can also be understood as the future state of society derived from future environmental constraints described in the steps of lifestyle design above. While placing yourself in this future state of society, you envision what a truly wholesome and fulfilling lifestyle would be like. This is an approach which is completely different from conventional extrapolation from the present.

Let us think, for example, of how to take a bath in 2030. We will go into more detail in Chap. 12, but let us think of a tub in which you fill with 300 L water, warming it to 40 °C. If every Japanese household takes one tub bath a day (in Japan, you wash outside the bath and the whole family uses the same 300 L), the volume of water required is equivalent to 37,500 of 25 m swimming pools. In 2030, finding energy enough to heat all this water will be extremely difficult. So, how can people then bathe? When this question is asked, many people will respond in the following way: "Showering only", "reducing the frequency of bathing", "only wiping your body with a cloth", "going to a nearby river", or "many people bathing together"—but are these enjoyable solutions to the problem? Does this future excite people and allow them to live wholesome lives? Unfortunately, these options are neither fun nor exciting. Why, then, do people come up with such responses? What we are witnessing is a typical solution based on thinking in the forecasting mode. Our way of thinking, unfortunately, does not allow us to think about the two mountains of the global environment and prosperous living at the same time. When preference is given to the environmental constraints, people will think first of how to endure or put up with less, and we see in responses such as the above-mentioned. "Having many people bathe together" may, perhaps, be said to include a certain degree of enjoyment, but in any base, it is a partial solution to the problem posed. Water and energy conservation, fundamentally, are ideas born from thinking in the forecasting mode.

How would solutions then look when thinking in a backcasting mode? One might get a response such as, "take a bath every day, but one that doesn't use water." In a system of nature technology, one looks for the answer in nature, and the result is using 3 L of water to generate bubbles at a temperature of 70 °C to fill the tub for bathing. If you only use 3 L of water, the tub itself does not have to be very sturdy, and being light you can bathe anywhere in the house. You may even say that thanks to the environmental constraints, a new, wholesome lifestyle that you had not even thought of before comes into light. With the same kind of backcasting approach, an air conditioning system without a power supply that uses earth to cool has come into being.

When mentioning products such as "an air conditioner without a power supply" or "a bath that needs no water", many people for an instant look baffled. This is because they are trapped by conventional thinking which says that "an air conditioner operates on electricity and is installed inside the house" and are thus unable to even imagine an air conditioner without a power supply. An air conditioner without a power supply refers to wall or floor material that can regulate the humidity in the room. This regulation of humidity makes for air conditioning without the need for electric power. Something similar is the case with "a bath that needs no water." Most people share the stereotypical view that "taking a bath means filling the tub with hot water" and thus become entirely unable to imagine one without water. As people are accustomed to using things in daily life, a fixed way of thinking comes to dominate in which material items such as "an air conditioner" or "a bath" come to mind. There is thus an ingrained inability to think of actions or functions such as "regulating the air", "cleansing the body" or "warming the body." If people were able to acknowledge air conditioning or bathing as functions, they would probably not to be puzzled by expressions such as "an air conditioner without a power supply" or "a bath without water." In everyday life, people's way of thinking tends to center around things and appliances instead of focusing on more generic concepts describing action and function. This is also a result of thinking in the forecasting mode.

We also often have the feeling that something is wrong with the many visions of the future presented in various forums. Many such visions are under the strong influence of the forecasting mentality, often expressed with already existing technologies—only with more functions than today or in miniaturized or enlarged versions—as their point of departure. Buildings are imagined taller than today (enlarged), virtual communications between people increases (higher functionality), there are fewer trees and the cityscape looks void of life (higher functionality). Futuristic images are expressed as new shapes of buildings (generally, more round buildings), and humanity's age old dreams, such as being able to fly freely through the cityscape, are imagined to have come true.

5.6   What Is a Wholesome, Fulfilling Lifestyle?

What kind of lifestyle is a wholesome, fulfilling lifestyle? The wholesome, fulfilling lifestyle that people desire can be redefined as "a lifestyle which people, even under severe environmental constraints, find exciting and would like to pursue". If we can clarify what kind of lifestyle people desire, we may be able to find important cues to define a wholesome, fulfilling way of living.

We have written out 30 original examples of lifestyles generated with the lifestyle design methodology and used these to create evaluation aspects based on an evaluation grid technique (Fig. 4.1). Participants in this experiment are shown 30 different types of lifestyles and sort these into "lifestyles I would like to live" and "lifestyles I would not like to live." After that, they are interviewed about the reasons for their responses. From this, we extracted 340 phrases and, using the KJ method, reduced these to 55 different elements. After that, by integrating similar phrases, we finally came down to 40 different "constituting elements of a wholesome, fulfilling lifestyle" (Fig. 5.4).

The elements forming what people see as a wholesome, fulfilling lifestyle both in quantitative and qualitative terms elements express "richness." Undoubtedly, a lifestyle

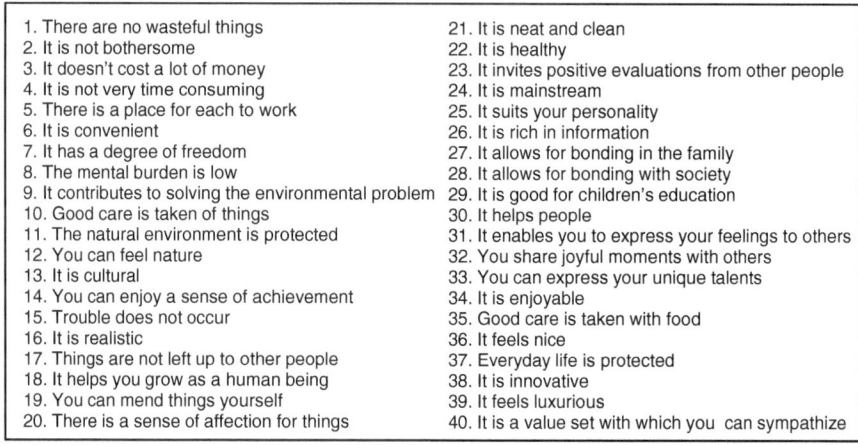

1. There are no wasteful things
2. It is not bothersome
3. It doesn't cost a lot of money
4. It is not very time consuming
5. There is a place for each to work
6. It is convenient
7. It has a degree of freedom
8. The mental burden is low
9. It contributes to solving the environmental problem
10. Good care is taken of things
11. The natural environment is protected
12. You can feel nature
13. It is cultural
14. You can enjoy a sense of achievement
15. Trouble does not occur
16. It is realistic
17. Things are not left up to other people
18. It helps you grow as a human being
19. You can mend things yourself
20. There is a sense of affection for things

21. It is neat and clean
22. It is healthy
23. It invites positive evaluations from other people
24. It is mainstream
25. It suits your personality
26. It is rich in information
27. It allows for bonding in the family
28. It allows for bonding with society
29. It is good for children's education
30. It helps people
31. It enables you to express your feelings to others
32. You share joyful moments with others
33. You can express your unique talents
34. It is enjoyable
35. Good care is taken with food
36. It feels nice
37. Everyday life is protected
38. It is innovative
39. It feels luxurious
40. It is a value set with which you can sympathize

Fig. 5.4 Constituting elements of a wholesome lifestyle

containing many of these elements is also a lifestyle desired by many people. Which elements are seen as more important, though, varies from person to person.

Figure 5.5 is a depiction of how 100 participants in a questionnaire rated the 40 elements for each of the four different lifestyle examples described above (Living in accordance with the seasons, shared batteries, the talking refrigerator, the purifying box). Participants were asked to evaluate to which degree these four lifestyles, on a scale from 1 to 6, applied to/contributed to each of the 40 constituting elements. The average score indicating the degree to which a lifestyle applied to (contributed to) the elements were 3.9 for "living in accordance with the seasons", 3.8 for "shared batteries", 3.1 for "the talking refrigerator", and 3.2 for "the purifying box." Looking at the average, there is virtually no difference in the degree to which "living in accordance with the seasons" and "shared batteries" are seen to apply to the 40 elements. In the same way, there is little difference in the average rating of "the talking refrigerator" and "the purifying box." When, however, one looks at the rating of each individual element, it becomes clear that the degree to which the different lifestyles are seen to apply differs greatly; each lifestyle has its distinct contour.

The social acceptability of the lifestyles was 76 % for "living in accordance with the seasons", 70 % for "shared batteries", 26 % for "the talking refrigerator", and 28 % for "the purifying box." This shows that lifestyles that were highly rated for the 40 constituting elements were also those most desired by participants in the experiment.

What makes for a wholesome, fulfilling lifestyle is greatly influenced by the structure or composition of the lifestyle (as expressed with the 40 constituting elements). Further analysis is needed to clarify the relationship between the structure (composition) of a lifestyle and the desirability thereof. If designed lifestyles are not accepted by citizens in society, the whole effort makes no point. An important task for the future is how to envision lifestyles with a high degree of social acceptability.

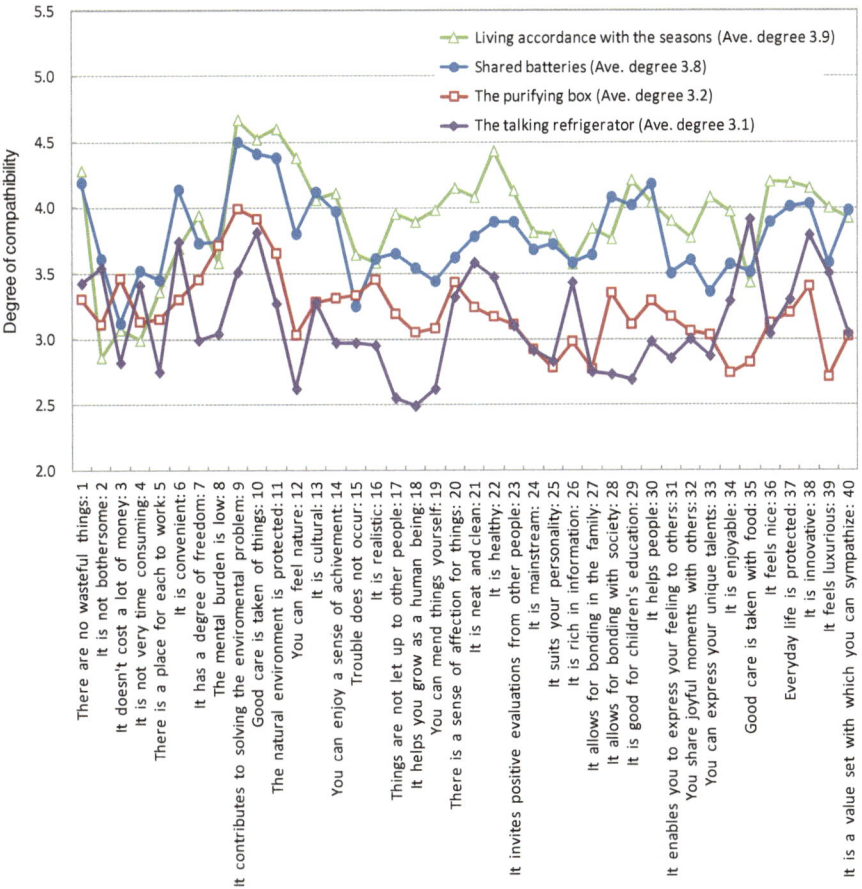

Fig. 5.5 The structure (composition) of lifestyles (2030)

Furthermore, it is necessary to check whether the lifestyle envisioned is not merely a partial solution approach. It makes little sense if the aspect of wholesomeness and spiritual fulfillment is so strong that the lifestyle ends up creating eco-dilemmas. To counter this, corporations, before bringing new products to the market, should examine whether these do not cause eco-dilemma-like phenomena.

Bibliography

Anderson KL (2001) Reconciling the electricity industry with sustainable development: backcasting a strategic alternative. Futures 33:607–623

Dreborg KH (1996) Essence of backcasting. Futures 28:813–828

Ezio M (2003) Sustainable everyday. Edizioni Ambienti, Milano

Green K, Vergragt P (2002) Towards sustainable households: a methodology for developing sustainable technological and social innovations. Futures 34:381–400

Hojer M, Mattsson L-G (2000) Determinism and backcasting in future studies. Futures 32:613–634

Holmberg JB (1998) A natural step in operationalising sustainable development. Greener Manag Int 23:30–51

Ishida H, Furukawa R, Maeda H (2008) Nature Technology Creation System. Kikai no kenkyu Yokendo Shuppan Tokyo 6(60):619–626

Lovins A (1976) Energy strategy, the road not taken? Foreign Aff 55:63–96

Lovins A (1977) Soft energy paths: toward a durable peace. Friends of the Earth/Ballinger Publishing Company, Cambridge

Masuda T, Furukuwa R, Ishida H (2013) Kankyou gijutsu senryaku ritsuantsuuru toshite no raifu-sutairu hazaado mappu (Lifestyle hazard maps as a tool forstrategic planning of environmental technology). Kankyokeizai seisakukenkyuu (Environmental Economics Policy Studies) 6(1):54–66

Masui T, Matsuoka Y, Hibino G (2007) Bakkukyasutingu niyoru datuonndannka shakai jitugen no taisakukeiro (Realization of anti-warmer society by backcasting) Chikyu kankyo 12:161–169

Quist J, Knot M, Young W, Green K, Vergragt P (2001) Strategies towards sustainable households using stakeholder workshops and scenarios. Int J Sustain Dev 4:75–89

Robinson J (1982) Energy backcasting. Energy Policy 10:337–344

Robinson JB (1988) Unlearning and backcasting, rethinking some of the questions we ask about the future. Technol Forecast Soc Change 33(120):325–338

Robinson J (1990) Future under glass: a recipe for people who hate to predict. Futures 22:820–843

Vergragt PJ, Jansen L (1993) Sustainable technological development: the making of a long-term oriented technology programme. Proj Apprais 8:134–140

Vergragt P, van der Wel M (1998) Back-casting: an example of sustainable washing. In: Roome N (ed) Sustainable strategies for industry. Island, Washington, DC, pp 171–184

Weaver P, Jansen L, van Grootveld G, van Spiegel E, Vergragt P (2000) Sustainable technology development. Greenleaf Publishers, Sheffield

Chapter 6
Lifestyles Envisioned with Backcasting

Abstract Of the lifestyles we outlined using backcasting, those in which nature was present in people's lives and where there was a sense of belonging to society had the greatest degree of social acceptability. Even when there is a slight sense of inconvenience, this is exceeded by the multitude of joys experienced in such a way of living. When we do cluster analysis of the lifestyles with high social acceptability, we find that factors such as "nature", "enjoyment", "self-growth", and "a sense of belonging to society" are included, indicating that these are elements that people, along with "convenience", unconsciously seek in their lives. Just as if they were yearning for childhood experiences—the good old days—people today unconsciously desire nature and enjoyment as much as convenience, and this could very well be one of the best motivations to start creating new lifestyles.

While people have this kind of latent desire, it is clear that the awareness of the degree to which environmental degradation has progressed is not sufficient. We fear that a pursuit of nature or enjoyment that fail to properly take into account the one-planet constraints we live under, may end up being very similar to visions of the future derived from forecasting.

Keywords Cluster analysis of lifestyles • Latent desires in lifestyles • Lifestyles in 2030 • The social acceptability of lifestyles • The structure of lifestyles

6.1 Lifestyles in 2030

Even today, if all people on Earth were to live the same lifestyle as the Japanese, we would need 2.3 planets. If, in such a situation, we merely continue imaging future lifestyles in an expansionist manner using forecasting, and base innovation in society and business on this, we will remain unable to overcome the barriers on the road to a society with low environmental impact and will further fan environmental degradation. Backcasting, on the other hand, works with the reality of "one planet"

E.H. Ishida and R. Furukawa, *Nature Technology: Creating a Fresh Approach to Technology and Lifestyle*, DOI 10.1007/978-4-431-54613-9_6, © Springer Japan 2013

Fig. 6.1 Image of a 2030 lifestyle

as the basis for contemplating wholesome, fulfilling lifestyles under the environmental constraints likely in 2030.

In the more than 1,500 possible lifestyles of 2030 we outlined so far using backcasting, we find exciting visions of a future world which would be unimaginable with thinking based on forecasting. Here, we would like to focus on 50 such examples of lifestyles that we designed in a joint research project with the Japanese advertising company Dentsu's Grand Design Laboratory in 2010.

The overall characteristics of the lifestyles described included a sense of nature in people's lives, protection of the natural environment, strong bonds with society, and culture as an element of everyday life. Even if there is a slight sense of inconvenience, the joys in life exceed this, people find themselves taking good care of things, and they live in such a way that continuous self-growth is possible. These are also innovative ways of living. In Fig. 6.1, we have created an image of what such a lifestyle might look like.

In the lower left corner, you will notice something resembling a vending machine in the bicycle parking space. This is a device for sharing community energy envisioned as part of the new lifestyle. Along the roadside, solar panels and storage batteries for the energy generated have been installed, and people in daily life share renewable energy in the community. Next to the bicycle power stand, a person is looking for something using a small portable device. His portable device has run low on energy, and he is looking for a place to recharge it. Accessing a website which records the remaining energy levels of people's portable devices, he immediately knows where the nearest recharging spot is, how much energy remains, and how many people are in the process of recharging at each spot. There is no need to stand in line waiting for a long time.

Slightly further up along the road to the right, a simple, one-passenger mobility device has been depicted. This device replaces the bicycle and drives around smoothly on electric power. When the landscape is flat, people in this community most often walk, but where there are ups and downs, they borrow they power of energy. The roads are curvy and are clearly no longer designed primarily for automobiles. The trees and other vegetation along the roadside are a joy for people to watch. Where the roads curve, both the wind and people tend to gather.

On the roof of the dome-shaped house in the center of the picture, small wind generators have been installed. Such micro-generators, which, based on the mechanism of a dragonfly's wing, revolve even in the slightest breeze, are installed here and there in the community, and the inhabitants are easily able to utilize even weak sources of energy that were not used before. Renewable energy has become a part of people's lives. Other small sources of energy are also being utilized. The energy from water running in small streams is harnessed in small scale hydropower, using the mechanism of the fin of a fish, and the cotton-bud like sticks on the dome-shaped roof are used to extract moisture from the air.

In the lower right corner of the picture, we see an area with small shops. There you can also rent a kitchen to have a neighborhood barbecue. Next to this, a man is cutting and cleaning a fish as a performance. In the shops, many things are sold by weight so you can buy exactly as much as you need.

Above the shop area, there is a fashionable restaurant. You pick your own vegetables for cooking by the chef, and this allows you not only to relish the freshest possible food, but also to enjoy the experience of picking vegetables yourself. Also, seedlings for making herb medicines are sold. A new product has been launched—medicine that you buy as seedlings and grow in your own home. If you grow the medicine with care, this is more economical than buying medicine at the pharmacy each time you need it. In this restaurant, geothermal heat is used to grow vegetables underground, and you can also buy vegetables harvested here.

At the apartment building further back in the picture, the staircases are green as the inhabitants enjoy growing kitchen gardens up along the stairs. In a shared housing complex like this, there are many benefits to gain from sharing instead of owning, and sharing is becoming increasingly popular. We also see people who are enjoying dyeing simple clothing. Since the inhabitants jointly bought washing machines with dye kits installed, people have become very engaged in the act of washing their clothes.

Left of center in the picture, a large tree stands as a symbol of the town. The children use the tree as a hideout and here find enjoyable ways to play in nature. In the surrounding park, many fruit trees have been planted that people can freely pick and eat as they like. In the park, a toilet has been installed which doesn't use water. Near the park is a so-called "silver nursery", where the old people of the town help looking after the children. The old people pass on the wisdom they possess to their grandchildren's generation—how to live in balance with nature, old skills and fairy tales, and lessons about the natural way for people to interact and the social rules

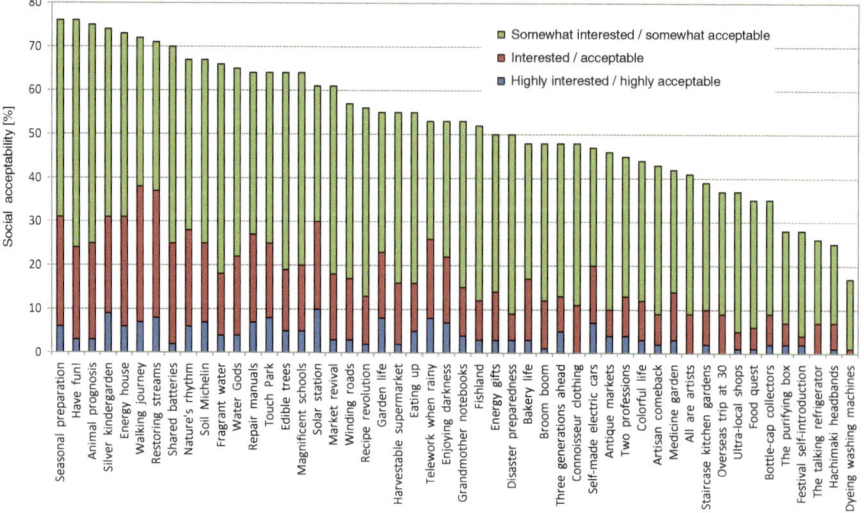

Fig. 6.2   The social acceptance of 50 different 2030 lifestyles (1,000 questionnaire from 20s to 60s)

supporting such interactions. The children love hearing such stories of the past from their wise old grandparents.

Further up along the hill, a family is enjoying work in their vegetable garden. Several shops are selling clothes that are both popular and have unique features. In these shops, conversations take place about how a piece of clothing made in a particular region is very durable, or how one from another region feels soft against your skin. People choose the cloth of their clothing much in the same way as a connoisseur picks a good wine. In a shop further back, there is a smart salesman famous in this neighborhood. When customers take too long to decide, he often purposely discourages them to buy: "If you are in such doubt, you don't have to buy that hat. You will do well enough with a piece of cloth around your head", he might say. A salesman, who is too busy to deal with indecisive customers, doesn't try to sell too hard. Based on his strong confidence, it is almost as if he is saying, "I am not interested in people who cannot instantly identify quality products". This kind of spirited exchange helps customers develop a discerning eye and raises the quality of products on sale.

Even further back in the picture, we find the most local of local shops. Here, you can enjoy a meal that nobody but this shop's chef could possibly prepare. This shop is not to be found in any guide books, and the local residents use it as a kind of hideout where they know a delicious meal is always available.

The above is an example of a lifestyle of 2030, designed with the backcasting method. In the joint research project, we wrote out 50 such examples of lifestyles, and analyzing these with a questionnaire-based survey, we were able to uncover the characteristics of new, designed lifestyles of 2030.

Figure 6.2 shows the result of an analysis of the social acceptability of the 50 different 2030 lifestyles we outlined in collaboration with Dentsu Grand Design

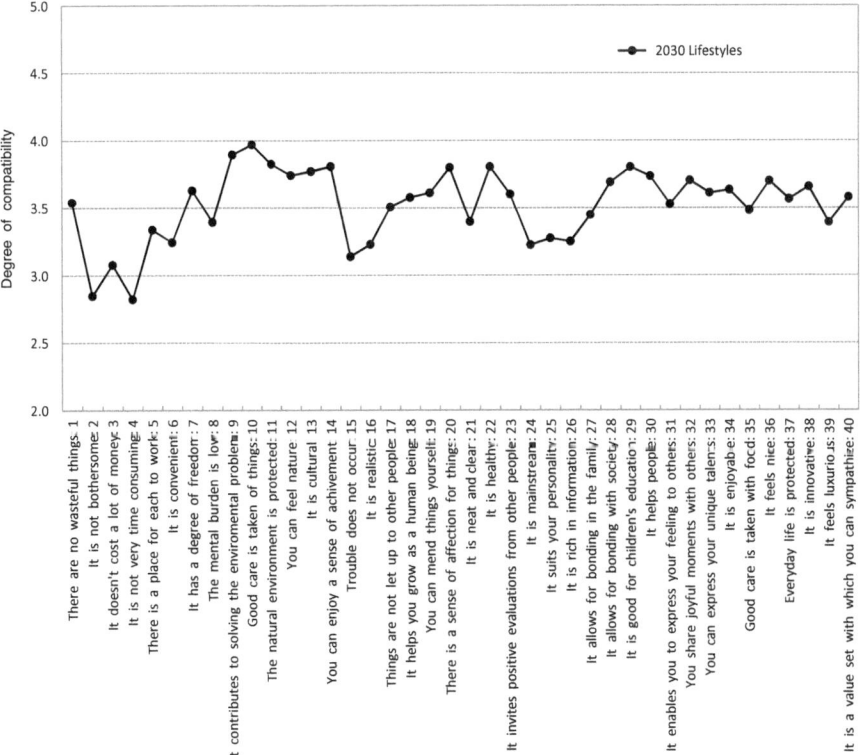

Fig. 6.3 The structure of 50 different 2030 lifestyles

Laboratory (Takido et al. 2010). The highest percentage of social acceptability was 76 %, and eight out of the 50 different lifestyles scored more than 70 %. The lifestyles designed with the backcasting method are somewhat like parallel worlds that one would be unable to discover with thinking based on forecasting. It is only by describing these lifestyles in concrete detail, that we become able to evaluate them. The average social acceptability rate was 51 %. This shows that lifestyles we could not have envisioned with forecasting are quite acceptable even with people's present values.

Can we perhaps say that such lifestyles exist, unconsciously, in the minds of today's citizens? So far, both consumers and engineers have been trapped by conventional forecasting, and thus lifestyles that were actually—albeit unconsciously—desired by people have not been expressed or pursued. With such desires hidden deep in people's minds, technological development in Japan continues to progress towards the realization of other types of lifestyles derived from roadmaps created with forecasting. Let us take a closer look at what it really is people are seeking.

Figure 6.3 is a graph indicating to what degree the 50 different lifestyles, on average, apply to or contribute to the 40 evaluation elements described in the previous chapter (using the lifestyle evaluation elements derived from the evaluation grid methodology). On the X-axis, we see the 40 different evaluation elements, on the Y-axis, a score from 6 (highest degree of compatibility) to 1 (lowest degree of compatibility). The median value for compatibility (the degree to which a lifestyles applies/contributes to an element) is thus 3.5.

An overall trend emerging from this plotting of the 50 different 2030 lifestyles is that they are evaluated to be somewhat inconvenient, requiring effort, money and time. On the other hand, they are also regarded to be lifestyles that contribute to the solution of environmental problems and that provide people a life in touch with nature. Also, the lifestyles are seen to incorporate good care taken of things and a sense affection for items you use in everyday life. Scores are also high for items indicating strong bonds with society, beneficial to children's education, and helpful to other people. Items relating to enjoyment in life also score slightly higher than the average.

6.2 The Measurement of Social Acceptability and Contributing Factors to Raising It

To what extent are these 50 lifestyles accepted within people's present value systems, and what factors contribute to the degree of acceptance? Let us dig deeper into the measurement of social acceptability and the unconscious desires of people hidden therein. Using the 40 constituting elements of lifestyles as a measurement, let us here try to make a quantitative analysis of which factors play the most important roles in raising social acceptability. The 50 different 2030 lifestyles were designed through group discussions with a total of 12 people—six of each gender. The 50 lifestyles all depict different scenes, but in Fig. 6.4 we have clustered lifestyles with similar elements using cluster analysis and look at the relationship between the composite factors of each cluster and social acceptability.

Next, as shown in Fig. 6.5, we did a factor analysis for each cluster and arranged them in order of social acceptability. We named each of the factors contributing to

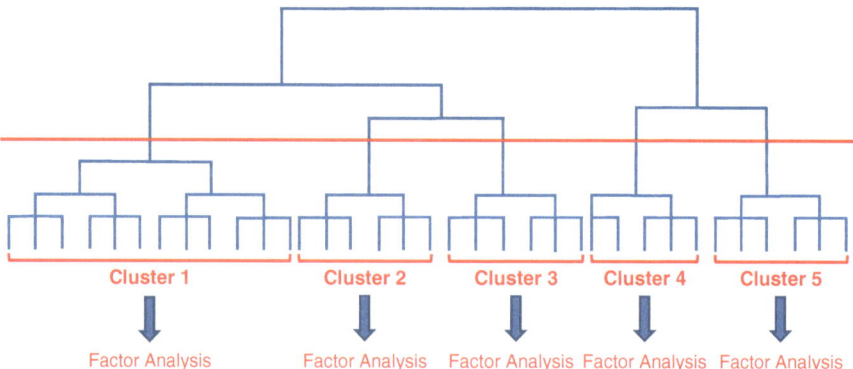

Fig. 6.4 An image of cluster analysis

Cluster no. (Order of preference)	Factor 1			Factor 2			Factor 3			Factor 4			Factor 5			Factor 6			Social acceptability
Cluster 1	Social belonging (A sense of belonging to society)			Nature			Self-growth			Convenience			Comfort			Nature/health			63.6
	27	30	36	11	9	12	18	17	16	4	2	3	33	35	8	21	12	35	
	3.85			4.18			4.01			3.3			3.93			4.2			
Cluster 2	Nature and things are both important			Enjoyable resonance (sympathy)			Convenience			Cleanliness/newness			Self-growth						57.3
	9	10	11	33	31	32	4	2	6	20	37	23	18	17	16			·	
	3.92			3.69			3.34			3.59			3.56						
Cluster 3	Enjoyable and luxurious			Nature			Convenience			Self/affection									52.4
	33	38	40	11	9	12	4	2	3	18	19	17			·			·	
	3.49			3.79			3.1			3.62									
Cluster 4	Comfort together			Convenience			Nature and things are both important			Affection together			Work						38.7
	33	35	31	4	2	3	11	12	10	19	18	30	5	27	29			·	
	3.38			2.94			3.08			3.36			3.4						
Cluster 5	Self-growth			Cleanliness/health			Nature and things are both important			Convenience			Enjoyable resonance (sympathy)						32.8
	17	18	16	20	21	38	9	10	11	4	2	3	33	32	31			·	
	3.06			1.89			3.34			3.25			3.08						

Fig. 6.5 Composite factors of lifestyles and social acceptability. Numbers are blue when subscale scores are higher than average. Numbers are red when subscale scores are lower than average (Takido et al. 2010)

social acceptability. For example, factor one of cluster one is called "A sense of belonging to society", the second factor "nature", the third "self-growth", the fourth "convenience", the fifth "comfort", and the sixth "nature/health".

From the cluster analysis, we found that in lifestyles with high social acceptability, factors such as "nature", "enjoyment", "self-growth", "a sense of belonging to society" were included, indicating that these are elements people unconsciously seek in their lives. Although "inconvenience" was also one of the factors included, we found that this did not impact greatly on the score of social acceptability, suggesting that people are willing to accept a certain degree of inconvenience in their lives.

With regard to lifestyles that showed a low percentage of social acceptability, we also tested whether adding the positive factors to these lifestyles would raise their acceptance rates. As a result, it became clear that by adding content which include or express positive factors to these lifestyles, it is indeed possible to raise social acceptability scores. Clearly, people, without being conscious thereof, yearn for lifestyles that include nature, a sense of belonging to society, self-growth and enjoyment.

6.3   The Structure of Trade-offs

In this way, the factors that most people, unconsciously, seek in their lifestyles were uncovered. In reality, however, there may not necessarily be a lifestyle with all positive factors included, and when people choose lifestyle, there will be situations when they have to trade one against another—choosing one positive factor while having to let go of another. When a mistake is made in this choice, the environmental impact of the lifestyle may end up increasing rather than decreasing. The factor

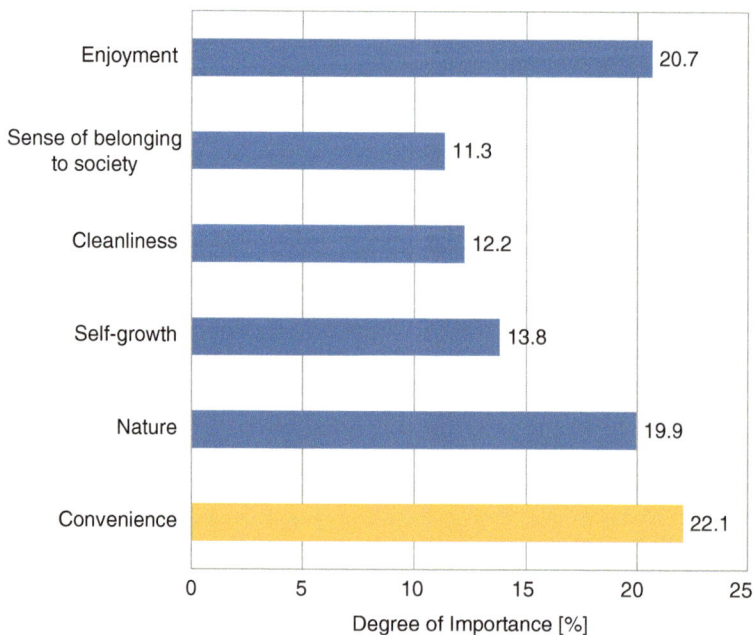

Fig. 6.6 Layers of dilemma and the importance of values (n=5,000) (Ogsta et al. 2011)

enjoyment is, in this respect, a particularly dangerous factor. If the enjoyment sought does not lead to increased environmental impact, there is no problem. If, however, it is of a wasteful nature leading to a greater burden on the environment, we have a problem on our hands. It is unlikely that the pursuit of nature in people's lives should lead to increased environmental impact, but when it comes to self-growth, there is a danger of greater environmental impact. Thus, all is well as long as the latent desires of people are well balanced, but people constantly face trade-offs and there is a hidden possibility of further burdening the environment in many lifestyle choices. Indeed, at the very moment when eco-dilemmas are born, there is a need to look carefully at the root cause, since we may well be seeing the outcome of a trade-off made between different factors.

What kind of relationship exists between the different, unconscious desires? Could people's unconscious desires not be one of the reasons for the existence of eco-dilemmas? In order to clarify what the structure of trade-offs looks like, we examined the factors constituting people's latent desires mentioned above using the so-called conjoint analysis (Fig. 6.6). An overall trend we discovered was that "convenience" was the factor people put the greatest emphasis on, with "nature" and "enjoyment"—both closely ranked—coming next. Then there was a large drop in terms of the emphasis put on "self-growth", "cleanliness" and "a sense of belonging to society". This clarifies the structure of the trade-offs people constantly make.

It is not clear how this structure of trade-offs came about, but it may very well be one of the reasons why people are unable to actually realize the kind of lifestyles (as depicted in the 2030 lifestyle exercise) they unconsciously desire. Although people yearn for nature in their lives, if this is done without taking environmental constraints into account and combined with the pursuit of convenience and wasteful pleasures, the future envisioned will be similar to one outlined with forecasting. If we can become able to combine our yearning for nature with the pursuit of convenience and enjoyment in respect of the limits of environmental constraints, we will move closer to the lifestyles needed in 2030.

Bibliography

Ishida H, Furukawa R, Dentsu Grand Design Laboratory (2010) Kimi ga otona ni naru koroni. Kankyou mo hito mo yutaka ni suru kurashi no katachi (By the time you grow up – a way of living that makes both the environment and people rich). Nikkan Kougyou Shimbun, Tokyo

Ogsta S, Furukawa R, Ishida H (2011) Kankyo hairyo kodo to kachikan no trade off kozo no kankei (The relationship between environmental behavior and a value judgment) General lecture summary from 26th annual academic meeting report of Research and Technology Planning Society, Japan, pp 405–408

Takido H, Furukawa R, Ishida H, Masuda T (2010) Kankyou seiyaku wo kouryo shita raifusutairu no hyouka kouzou chuushutsu to shakaitekijuyousei nikansuru bunseki (Analysis of the social acceptability and determination of evaluation structure of lifestyles taking environmental constraints into account). General lecture summary from the 25th annual academic meeting report of Research and Technology Planning Society, Japan, pp 436–439

Chapter 7
Pre-war Living in Co-existence with Nature

Abstract We have seen how people, unconsciously, seek "enjoyment" and "nature" just as strongly as "convenience" in their lives. What, then, is the nature of these two elements, "enjoyment" and "nature"? Numerous enjoyable things—games, the internet, movies—surround us in everyday life, but what people are yearning for is another form of enjoyment. Or, do people with "nature" mean that they are eager to go to the beach or hiking in the mountains? We do not believe this is the only form of nature people are seeking. In order to clarify the true meaning of "nature" and "enjoyment" in everyday life, we conducted a series of interviews with people around the age of 90 (below, nonagenarians) and made a qualitative analysis of the pre-war living in co-existence with nature. Through these interviews, we discovered more than 70 different pieces of wisdom or techniques of living in co-existence with nature. We found that "knowing that you are being kept alive by nature, knowing how to utilize nature, and knowing how to deal with the challenges nature puts forth" are like three origins of the Japanese way of living. These precious pieces of wisdom and techniques are the essence of a wisdom of life that the natural world in general possesses, which allows people to connect with nature, and which in structure resembles low impact technologies.

Keywords Community bonded by shared tasks • Interviews with nonagenarians • Living in balance with nature's rhythm • Roles in daily life

7.1 Pre-war Living in Co-existence with Nature

Throughout history, the Japanese have spent much longer time than most people today realize in a process of trial an error to master methods of living in harmony with nature. In the process of this, the Japanese arrived at the optimum way of living in the geographically limited and environmentally unique Japanese archipelago, and established regionally adapted ways to pass this knowledge on to the next generation. This valuable knowledge and wisdom, however, is etched into the brains of the older generation only, and thus is at risk of disappearing. The generations born after World War II became hooked on the merits of new convenient technologies imported from overseas and have since gone on to develop even more convenient things. Thanks to this, modern society is undoubtedly a highly convenient world in which to live. In exchange for this, however, many methods of living together with nature in wholesome, fulfilling ways are gradually being lost.

This deplorable state of affairs is not well known. The only people who have noticed are those who are around 90 years of age today, in whose memory the precious wisdom and techniques of living still exist. "Why doesn't the younger generation come to us for advice?", the nonagenarians lament. We have neglected the important task of passing on knowledge of wholesome, fulfilling lifestyles with a low environmental impact to the next generation. It is an urgent for us today to visit nonagenarians all over Japan and listen to what they have to say.

In a questionnaire survey we conducted in 2009, we found that 51.9 % of respondents are in regular contact with elderly people. In this respect, there still appears to be a mechanism to pass on stories from the past, but if you compare with pre-war Japan in which three generations lived under the same roof, one could also say that this possibility has halved. The result of the survey shows that people are concerned about the gradual loss of, and want to recapture, such elements of daily life as "being in touch with nature", "time that flows without paying attention to a watch", "face-to-face interaction with other people and social manners", "events of the four seasons", "traditional events", and "enjoyment". Although still only vaguely, a sense of premonition of the crisis to come is detectable in these responses.

The nonagenarians we interviewed grew up with the environment and education of pre-war Japan, and turned 20 before the war started. They had reached 40 years of age, and were central figures in Japanese households, in the early 1960s, when per household energy consumption was less than 60 % of what it is today. They know how to enjoy themselves even under severe constraints and possess the vitality that comes from being able to make most things yourself. Let us here introduce the reader to the recollections of one of these nonagenarians.

Recollections of nonagenarians—a 91-year old woman from Miyagi Prefecture (Interview conducted in Feb., 2010)

> When it was cold, in January or February of the old, lunar calendar, we made *shimi-dofu* (tofu that has been frozen and dried) before the New Year came. Two or three households made *tofu* together, froze it outside and tied strings around it before it was dried. We also made *natto* (fermented soy beans) together. We boiled soy beans, wrapped and tied them up in straw, and then folded many of these up in a woven straw mats. These packs were then put in a container with rice husks, which we had plenty of in those days, and we let it ferment with a quern on top as a weight. After that, the rice husks and straw were used as compost. We made *miso* and soy sauce, too. To make soy sauce, we first steamed wheat, mixed it with boiled soy beans, salt and water, and then let it ferment. We stirred this mixture once or twice a day so that mould did not form. Potatoes can last a long time, right?
>
> There were always leeks in the fields, and sweet potatoes put in earth could last for quite some time. We made pickled cucumbers with lots of salt, stored them in earth in a shaded place and then washed the salt off before eating them. We sure did eat a lot of things in

season. A fish monger came to our house to sell fish, and since some of our relatives lived in the coastal towns of Shiogama and Yuriage, we were often given fish as well. Bonito, sardin, saury, salmon…when we received a large amount, we would share the fish with people in the neighbourhood, and we stored some of it pickled with miso, salted or dried. We were a big family, so no matter how much we had it disappeared quickly. We used to say that animals with four legs were not for eating, but instead we ate the chicken that we were raising. When you pour boiling water over the dead chicken, it is easy to pull off the wings. I couldn't stand the smell of this, though. For breakfast we had white rice, *natto*, *miso* soup and some leftovers from the night before. For lunch we had some boiled vegetables and salted food. For supper, some fish was added to this. At New Year, we did not offer sweet bean paste to the gods, but we mixed soybean flour into sugared water to make ricecakes; and then we also had *natto* rice cakes and a special soup with vegetables and fish. On the second day of the new year, about 30 of our relatives would gather for a New Year celebration. We would have *kombu* seaweed with squid, a vegetable mix with carrots, fried burdock root (*gobo*), vinegared *daikon* and cod, *harako* fish eggs, and fish of the season. Our parents bought us a *kimono* or wooden *geta* sandals for the New Year. For the midsummer Tanabata celebrations (July 7th), we made special decorations outside, and the children would sit on bamboo mats under these having a special supper before it got dark. It was really lots of fun. In those days, there was no underwear, so we just wore a loin cloth and went playing in the river. There was one spot where the farmers had dammed the water to diverge it into the rice paddies, and in summer we would always be playing there. At the end of July, before the traditional *Obon* celebration (when the souls of the deceased were said to return to their families), we would remove the *tatami* rice straw mats from the floors, hang them up to be dried, and clean the wooden floors. The rice mats would be returned in September, before we started getting busy with the rice harvest. When we had finished mopping the wooden floors, next would be the cleaning and repair of our *futons* and bedclothes. We were quite busy, you see. We took out the cotton inside the futons to be washed (and if the cotton had become too hard through use, we would beat it to make it soft again). Laundry was washed at the river using a laundry board. In the summer, we used the water destined for the rice paddies, so it was pure and clean.

Drainage from the kitchen was, I believe, used as fertilizer in the fields. To gain fuel, five people or so would together harvest wood from one mountain. With hatchets and saws we went into the mountain, cut branches off tree trunks, and carried the lumber down by horse- cart. During the war, there were no men around, so I also had to go. It was hard work. In winter, we kept ourselves warm with heated *kotatsu* tables or around the open fireplace. Our grandparents used hot water bags to stay warm. The wives of men born in the family were not let into the warmth. They had to endure the cold. For kibe and frostbite, we rubbed on a plant called snake gourd found in the cedar forest. Kibe hurt, so we rubbed on boiled grains of rice and put pieces of cloth over this. My grandfather would spray *shochu* liquor which had been preserved with a viper (snake) inside on the wound. When I found this strange concoction in the cupboard, it really shocked me. I was so scared I couldn't even look at it. For a snack, we had dried rice cake that had either been deep-fried in oil or fried in a clay pot. We also got rice snacks called *pongashi*. When I was young, one of the things I looked forward to was going to Mount Kogota in spring to pay homage at a shrine. Wives from our neighbourhood walked together to Sendai Station, from where we took the train. We enjoyed Kogota sweet buns. It was a one-day trip. To wash our bodies or hair, we used the fruit from the *saikachi* tree which could be made into foam to use for washing. We threaded the fruits onto wires and used them. Before the war we didn't wear make up. For skin lotion, we used a home- made extract from the plant called sponge cucumber—although that was quite a bit later. As children, we didn't brush our teeth. When we starting brushing them, it was with salt.

The way of living described here includes enjoyment, expresses vigour, and, for people of the present day, may appear to have many attractive aspects. This is because it is full of elements of life that no longer exist in Japan and numerous activities that have become lost after the war, without people noticing. Facets of life that have been passed on from generation to generation, and that we feel should not be lost, still reside somewhere deep inside the Japanese, and when we hear people talking of this, it somehow tugs at our heartstrings.

7.2 Interviews with Old People (Nonagenarians) as a New Methodology

Industry, bureaucrats and scholars are doing their utmost to find solutions to the global environmental problem. Are, there, however, anyone who is listening carefully to what old people, such as nonagenarians, have to tell us? How many people have noticed that listening to nonagenarians may be a shortcut in our search for the elements required to create a sustainable society? Unnoticed by most, knowledge of pre-war living, which had the low environmental impact required in a sustainable society, is gradually being lost as the memories of nonagenarians fade. People in modern society are bustling about trying to obtain a convenient life, but we need to stop and contemplate what is being lost and to record such important memories before they vanish entirely. We should apply the wisdom and techniques of life of the nonagenarians, in a modernized form, to our present day lives thus creating new future ways of living. Utilizing such wisdom and techniques in designing a plausible future, we must advance efforts to reform lifestyles before global environmental problems reach a critical threshold.

Interviews with nonagenarians are not simply about listening to and writing down conversations with old people. Deliberately, we search for elements and knowledge of living that is different from the present day. These are things that have been lost. "Living in harmony with nature's rhythm", for example, is one element of life that has been lost. In the old days, even living in a city, people would take good care of nature, co-exist with nature and utilize nature in everyday life. People began their daily activities as the sun rose and birds started chirping, and ended them as the sun set. A daily routine in harmony with the movements of the sun were rhythmical and pleasant. When spring arrived, preparations would be made in the fields, and when autumn came, the fields would be harvested. On the other hand, people would feel anxious if the cherry blossoms did not occur at the normal time of year. People would be calmed by the rhythmical sound of chirping insects in the late afternoon and reflect on the day gone by. It was comforting to live in balance with both the short and the long cycles and rhythms of nature. Today, however, such a way of living has been almost entirely lost.

"A community tied together by shared tasks and mutual support" is another element that has been lost. In the past, people in a community would share and help each other in important tasks thus strengthening mutual relationships. Basically, people would share tasks in large, multi-generation families and everyone had his or her role in daily life, but even in the largest of families, there were things that could not effectively be done. It was crucial to share and support each other in the community when it came to, for example, the repair of a house roofs, the ownership and use of agricultural equipment, or the protection of important water sources. Bonds in the community were strengthened by the sharing of such important tasks. This way of living has, however, has almost become a page in a history book.

The interviews with nonagenarians are, thus, a new methodology to learn about ways of living that are gradually being lost (but that we must not lose) in modern society.

7.3   Ways of Living That Are Disappearing

Between 2009 and 2011, a research group led by the authors of this book conducted interviews with more than 65 nonagenarians living in Miyagi Prefecture. The interviews were written down and the team spent much time thoroughly reading through these records. Through an analysis of these interviews, we extracted more than 70 different ways of living or values, including the above-mentioned "living in harmony with nature's rhythm" and "a community tied together by shared tasks and mutual support" (Fig. 7.1).

After the Great East Japan Earthquake, we have visited many evacuation shelters. Among these shelters, we found some to be in such a gloomy mood that we hardly dared to open the door, while others—even in a situation where people had lost everything—were filled with the smiles and laughter of the inhabitants. In shelters with a positive spirit, some people had linked pipes found among the debris together to transport water to the shelter from mountains kilometers away, some had created makeshift water basins to hold large fish that had been washed up by the tsunami, and yet others had made places to store water for drinking, cooking, or for doing the laundry. The elderly were giving instructions to the young, and the women, while chatting and joking, shared the task of cooking for the inhabitants. Even the children had a role to play in the daily life at the shelter. As we visited several times, it gradually turned spring and some of the inhabitants—who supposedly had lost all in their lives—even gave us butterbur pickled with *miso* as a present to take home. When we carefully observed these energetic and cheerful shelters, we learned that the 70 keywords of living extracted from our interviews with nonagenarians were very much alive even today. And, we now feel quite confident that these keywords are elements of culture and life that we, at least in Japan, must not allow ourselves to lose.

Many of the values expressed in these keywords are, however, gradually disappearing in modern society. We believe that many people today feel a degree of nostalgia for these values. Is this not because these are ways of living essential to creating a sustainable society that the Japanese have cultivated through centuries?

Let us try to look at some of these gradually disappearing aspects of living, comparing pre-war living with that of today.

Nature
1. The pleasure of living in balance with nature's rhythm
2. Reading the signs of nature
3. Utilizing nature in daily life
4. Being prepared for natural disasters
5. Proximity to other living things
6. Playing with other living things

Community/sharing
7. Helping each other with protection of water sources, agriculture, the making of roofs
8. Community is bonded by shared tasks and mutual support
9. Places and events provide a sense of belonging for people in the community (Shrines and temples, festivals)
10. Enjoying life in the community/region
11. Annual events are important
12. Sharing the mountains, fuel and water

Bonds in the family
13. Kindness in the family
14. Mutual support beyond the family
15. Living together with people who are not relatives is common
16. Passing on knowledge to the next generation through daily life
17. Children have a role/chores in the family
18. The elderly have a role/chores in the family
19. The head of the household has a role in the family
20. Work in the house and in the community
21. Children learn by their own
22. Children have their own world

Taking good care of things
23. Things used in everyday life are grown / nurtured and preserved with care
24. Maintenance (garden, tools, clothes)
25. No excess of things
26. Acting with half a year into the future in mind
27. Using up things, finding multiple uses, using over generations
28. Repairing and using again

Cycles
29. Food, fuel, timber etc. are procured and consumed locally
30. Circulating things (circular use)
31. Gathering things and food in the environment
32. A life of self-sufficiency

Growing and excelling
33. Snacks are grown in the garden and are part of the landscape
34. A useful garden
35. There are sheds and storage houses
36. The shape of the house reflects the way of living
37. The body is also a tool
38. You are creative and make things yourself
39. Songs as a part of life
40. There are sounds present in everyday life
41. A culture of metaphorical enjoyment

Making things
42. The house is a place of production
43. Visitors are entertained in the home
44. A working relationship with fire
45. Unique ideas to erase smells

Contact with the outside world
46. A lot of time is spent walking
47. Many different means of transportation
48. There are mechanisms and places encouraging people to meet
49. Going back and forth between the city and rural villages
50. Contact with the outside world

Business and the trades
51. An entertaining street of shops
52. Small scale trading
53. Specialized shops and craftsmen
54. Delivery businesses
55. Things are sold by weight
56. One person may have different trades to live by

Different values from today
57. Hard working
58. Changes of job, removals, changes in everyday life
59. Value that is not counted in money
60. A different perception of time
61. You create the community yourself

Gratitude
62. Water is used with care and gratitude
63. Gratitude is felt for things
64. Respect is shown for nature
65. Respect is shown for ancestors

Enjoyment
66. A different notion of luxury
67. A sense of laxness and generousity
68. Moderation – knowing just how much is enough
69. A clear distinction between the extraordinary/celebratory (*hare*) and the daily (*ke*)
70. Familiarity with both life and death

Fig. 7.1 Examples of pre-World War II ways of living

7.3.1   "The Pleasure of Living in Balance with Nature's Rhythm"

In pre-war living, people ate food in season and enjoyed moon viewing; they planted trees and plants which provided a sense of seasons in the garden, and generally found pleasure in living attuned to the rhythms of nature. Today, the rhythm of life is centered around humans, and when something is needed in daily life people immediately buy whatever it is. We have come to live in haste and impatience, and our lives have a high impact on the environment. Because the way we spend time in our lives is not in balance with nature's rhythm, we need otherwise unnecessary lighting, and because we live in an indoor environment which is inverse of the outdoor climate, we increase energy consumption to stay comfortable.

7.3.2   "Reading the Signs of Nature"

In pre-war living, people read the signs of nature and used them to decide how to behave. They watched the mountains to see how much snow was left in spring, listened to the sounds of waves coming from the sea or carried from afar by the wind, or they made judgments from the cold after a night of frost. This was, indeed, a way living together with nature.

7.3.3   "Wisdom to Utilize Nature in Daily Life"

In pre-war living, nature was wisely utilized in a number of ways. For instance, there was often a kind of mezzanine in houses designed so that sunlight would warm the space to prevent vegetables stored for the winter from freezing. As modernity progressed, we have become used to merely pushing a switch, while we have stopped using nature cleverly to avoid using energy.

7.3.4   "Being Prepared for Natural Disasters"

In pre-war living, most households kept a "starvation storage" (*gashikoi*) on wooden beams in the main house with rice packed in straw bags, typically some six large bags. Some households even kept 30 straw bags filled with rice, enough for 1 year's consumption. Branches were also stored in a wooden hut as a fuel reserve. Today, families no longer store this much food or fuel in preparation for natural calamities. When a disaster strikes, people now rely on others for help.

7.3.5 *"Proximity to Other Living Things"*

In pre-war living, people would catch crabs at the beach, and in marshes or rivers would find crucial carps, loaches, or sometimes even eels. Cows would be mooing, and chicken were running around the house. Some households even kept silkworms inside the house. Today, however, the fish in marshes and rivers have all but disappeared, and while some people do keep pets, there is a no longer a sense in daily life of other living things providing fulfillment of the heart or enriching people's lives. The distance to other living things has increased.

7.3.6 *"Community Is Bonded by Shared Tasks and Mutual Support"*

In pre-war living, families would get together to share tasks in the community that could not be carried out by one household alone. Agricultural equipment, for example, was co-owned, the cost of repairing roofs was set aside in advance in a kind of community trust, and repairs were undertaken jointly in a rotational system. In these and other ways, people supported each other mutually with essential tasks in everyday life. It was not like today, when people do everything individually in each household, leading to waste and high environmental impact.

7.3.7 *"Places and Events Providing a Sense of Belonging for People in the Community"*

In pre-war communities, there were places or events, such as Shinto shrines and traditional yearly events that anchored people and gave them a sense of belonging. When important markets were held, children would have a day off from school, and through the participation of all in the community, bonds were strengthened. Today, not only are there fewer places providing people in a community with a sense of belonging, but more and more people even avoid visiting community spaces for safety reasons.

7.3.8 *"Annual Events Are Important"*

In pre-war living, annual events were seen as important. When it was time for a festival, people would buy new clothes or wooden sandals, relatives would gather, the town would bustle with activity, and all in the community would have fun. At the *O-bon* festival (Festival of the Deceased) in summer, people would burn piles of firewood, and it was said that bathing in the smoke of this fire would help keep you healthy for the coming year. Such annual events were passed on with great care from generation to generation.

7.3.9   *"You Create the Community Yourself"*

In pre-war living, people were actively collaborating on issues important to the community. Groups of a few young men would gather, stay and cook together, hold memorial services for deceased who did not have any relatives, help with festivals, hold parties celebrating the completion of hard work in the community, and help with funerals. Community was created and maintained with the full participation of its inhabitants

7.3.10   *"Sharing the Mountains, Fuel and Water"*

In pre-war living, mountains, fuel and water were common property and were shared among members of the community. There were also cases when a few families together owned and shared fuel. To own all of this was simply too much for one family. The size of one single family was too small to ensure that the community could live in balance with nature. Therefore, nature was often treated as common property for shared usage.

7.3.11   *"Joint Effort and Mutual Support"*

In pre-war communities, there were self-help organizations called *Yui*, in which 20–30 people would get together and share work like, for instance, rice planting. To make *miso* paste, three people or so would share the work of boiling soy beans in a large iron pot, and those who did not have bath facilities at home, would visit other families in the neighborhood to bathe. In this way, it was part of daily routine to work together and support each other in the community.

7.3.12   *"Companionship with Other People as a Source of Vitality"*

In pre-war living, the source of energy and vitality for many people was companionship with others in the community. Men from trading families, for example, often had the custom of going together to take a morning bath. Pulling their two-wheeled carts with goods for trading was hard work, but these peddlers were invigorated by chatting with people while selling their vegetables or other wares. Many people also looked forward to the activities of young people's groups in the community. As urbanization has progressed, such companionship in the community is rarely seen.

7.3.13   "Living Together with People Who Are Not Relatives Is Common"

In pre-war communities, there were more members in the family than today, and in addition to this, relatives would often come to visit, or people from the neighbourhood would come by to chat or play, thus making the house a place bustling with activity. It was quite normal to have 10 or even 15 people in the house at any given time. Living together was a natural part of life. As the nuclear family gradually took over, the number of family members, first of all, fell, and neighbours no longer come by to visit very often. Common sensical rules for living together with people who are not family members is gradually being lost.

7.3.14   "Passing on Knowledge to the Next Generation Through Daily Life"

In pre-war living, people had a clear idea about what to pass on to the next generation, and through daily life, knowledge and techniques were communicated to the children by members of the family. Living under the same roof, adults by example showed children how to act, and even casual mutterings played an educational role. The dining and living rooms served as actual, visible spaces of living in which children again and again were impregnated with essential knowledge. Mechanisms for passing on wisdom across generations were built into everyday life.

7.3.15   "Everybody Has a Role"

In pre-war living, everyone, from children to adults, had a role to play in the family. Elder siblings looked after their younger brothers and sisters. From a very young age, all would be given chores in the house. Small, narrow spots where only a small hand could enter were cleaned by the children. As the children grew older, their chores would change. Sometimes, children would prioritize their chores at home higher than playing with their friends. In this way, everybody had a role in the household and taking part in the daily chores gave a sense of fulfillment.

7.3.16 *"Children Have Their Own World"*

In pre-war living, children had their own world unfolding. There were countless ways of playing in nature, such as gathering chestnuts on a mountainside. Children would bathe naked in the river, tease each other and have fun in numerous ways. This was a world in which the adults did not interfere. Often, parents did not know what their children were doing where, but when a dangerous situation occurred, people from the neighborhood would caution the children. There was a wide and exciting world for children.

7.3.17 *"The House Is a Place of Production"*

In pre-war living, the creation of value—production—was an everyday part of life in the household. In the house, silkworms would be kept, clothes were made, food was cooked, furniture was manufactured, ropes were weaved, and food was dried to allow for long term preservation. Little by little, however, the house has become a place for watching TV or eating; mainly a place of consumption.

7.3.18 *"Things Are Maintained"*

In pre-war living, it was normal to maintain things and when something broke, it was taken to a craftsman for repair. Tools necessary in daily life were repaired over and over again and used for a long time. Along with modernization, however, there are fewer things people are able to maintain themselves, and since it has become more economical to buy a new item than to repair it when broken, the use life of things has shortened. Because of this, people also find it more difficult to feel affection for daily items.

7.3.19 *"No Excess of Things"*

In pre-war living, there were only few things inside the house. When a woman got married and joined the husband's family, it is said that she only took one wooden washbasin, a wooden box for clothing, and bedclothes. In larger cities, people often moved and owned only few things. As modernization progressed, things of little use started cluttering people's houses, and this excess of things even stresses people today.

7.3.20   "Acting with Half a Year into the Future in Mind"

In pre-war living, people sometimes conducted their everyday activities with six months or more into the future in mind. The firewood used in winter was dried about 1 year before being ready. To make *haori* jackets for their children when married, people would start raising silkworms long before in preparation.

7.3.21   "Food, Fuel, and Timber Are Procured and Consumed Locally"

In pre-war living, food, fuel and timber gathered in hills behind the house or in nature nearby were used in everyday life. Fish and shellfish were caught in nearby rivers or the sea, and wood or cedar leaves were used as fuel. Most things used in daily life were produced and consumed locally.

7.3.22   "Circulating Things (Circular Use)"

In pre-war living, things were used up, or what was leftover was reused in a cyclical resource system. For example, ashes and dung was used as fertilizer, fish intestines were eaten, and oil that had been used for cooking was reused as fertilizer. Belts used when working were weaved with old ragged pieces of cloth. The residue leftover when boiling soy beans was used for washing clothes. Children wore clothes handed down from older siblings, and paper taken off the sliding paper doors in the house was reused for other purposes. Gradually, however, people starting throwing things away so frequently that the adjective "disposable" is now attached to much of what we use.

7.3.23   "A Useful Garden"

In pre-war living, edible trees were planted in the garden such as sweet and astringent persimmon, fig, loquat, and plum, and other plants were used to make tools. In this way, the garden was useful to the household. Fruit from the *saikichi* tree was used instead of soap.

7.3.24   "There Are Sheds and Storage Houses"

In pre-war living, a family would be using several sheds or storage houses. Among others, there were workshops (huts), storage sheds for rice and *miso*, bird houses, horse barns, storerooms, sheds for toilets and for bathing, and storages for ash fertilizer. There was good reason to have a separate shed for each distinct purpose. It may also, partly, have been for aesthetic reasons.

7.3.25   "The Shape of the House Reflects the Way of Living"

In pre-war living, the construction of a house reflected the way of living. If a family, for example, raised silkworms, the location of this activity was arranged in such a way that the heat from the *irori* open fireplace would be transferred. Toilets and bath facilities would be set up in another building. The porch was a cool place where people could relax while drinking a cup of green tea. With the arrival of mass-production and mass-consumption, however, people started fitting their lifestyle to the shape of the house rather than the other way round.

7.3.26   "Songs and Sounds as a Part of Life"

In pre-war living, songs were sung in the house or at the work place as part of the work routine. Songs sung while threading yarn were passed on from mother to child. Babies were nursed while singing, and there are records of annual events in which people walked singing around the community while receiving rice cakes from the houses visited. Apparently, this endeavor was both enjoyable to people and made them feel slightly embarrassed at the same time.

"Visitors Are Entertained in the Home"

In pre-war living, when a guest came to visit, the family would cook and entertain the visitor, be generous with the *sake*, and maybe ask the visitor to stay the night over. It was common to entertain visitors in the home. Some households even had an *irori* open fireplace especially for visitors.

7.3.27   "A Working Relationship with Fire"

In pre-war living, fire was often used inside the house. People would be used to washing large pots pitch black with soot, they would have experienced burns from fire and would always be alert not to cause a fire. Special effort would be made at the end of the day to avoid the fire extinguishing entirely in the course of the night. Today, people rarely use fire in the home—at the most a gas stove— and while convenience has gone up, we have forgotten how to deal with fire in everyday life.

7.3.28 "A Lot of Time Is Spent Walking"

In pre-war living, people often walked long distances. The children would often be asked to walk eight kilometers just to do a small errand, and in the daily trade people would often walk 20 km in a day. Everything was done walking. It was quite normal that it took two hours to get somewhere and another two to get home. This was so common that the phrase "walking is working" was used. However, as modern means of transportation have taken over, people may walk for health reasons, but there is no longer a need to walk as a means of getting from point a to b.

7.3.29 "There Are Mechanisms and Places Encouraging People to Meet"

In pre-war living, special organizations for young people of the age 15–20 were active in the community. In the agricultural off-season, the young in these groups would enjoy singing or dancing together. Sometimes, people would meet in such a group and later get married. Also, by doing errands walking even to destinations quite far away, people had the chance to meet people from other communities, which also sometimes led to marriages.

7.3.30 "Going Back and Forth Between the City and Rural Villages"

In pre-war living, some people often travelled between the city and rural villages as part of their trade. Fish or wooden fuel collected in a rural village would thus be transported to the city, and the trader would buy oil and other items only available in the city to take back. Most often, the trader had regular customers to whom he would sell his goods. After transportation infrastructure has been put in place, however, this movement of people between urban and rural areas has disappeared, information is no longer gathered from people, but from the newspaper, tv programs or the internet, and goods are not locally procured but can be purchased at a supermarket.

7.3.31 "An Entertaining Street of Shops"

In pre-war living, small shops lined the streets and craftsmen would be producing or repairing their goods with passers-by looking on. People would never get tired of watching the craftsmen at work. The shopping street thus also had an entertaining function. With the shift to mass-production, however, the manufacturing and repairing once carried out by craftsmen have become like a black box invisible to ordinary people, and thus the enjoyment has also vanished. Also, producers are no longer under the beneficial pressure of knowing they are being watched by consumers.

7.3.32   "Small Scale Trading Abounds"

In pre-war living, there were many different types of shops and each one of them was, generally speaking, of a small scale. One could find, for example, platemaking shops, masonries, shops selling hoops for barrels and baskets, plasterers, carpenter shops, tin(plate) manufacturers, barbers, shops selling wooden *geta* sandals, shops selling weaved bamboo sandals, horseshoe manufacturers, watchmakers, dyers, blacksmiths, *tofu* shops, fishmongers, shops selling *konjak* gel, soy sauce shops, Japanese sweet shops, cotton shops, public baths, Japanese lantern shops, printers, machine workshops, wood recyclers/resellers, and workshops making decorations and ornaments. This helped maintain a balance between supply and demand of a particular region.

7.3.33   "Delivery Businesses"

In pre-war living, there were many peddlers like, for example, fishmonger who would carry fish in shoulder baskets from coastal regions to sell in the towns. From far away, peddlers carrying fancy goods in bamboo baskets piled high would come to sell their stuff, and there were many different kinds of delivery businesses.

7.3.34   "Things Are Sold by Weight"

In pre war living, food at the grocery store was sold by weight. Even *tofu*, raw fish, soy sauce, and *sake* were sold by weight. Since people bought only as much as they needed, less food was thrown away due to excessive purchasing.

7.3.35   "Several Trades to Live by and Changes in Life"

In pre-war living, some people had several different occupations. For example, farmer and fisherman, or transporter and seller of vegetables at the same time. A work-style in which people worked until retirement in one company was not found.

7.3.36 *"Value That Is Not Counted in Money and the Mechanism of Money"*

In pre-war, bartering was a natural part of life. The value of an item was not necessarily counted in money, but was fixed mutually by the people bartering. A person might bring firewood and receive oil in return from the counterpart. Day labourers were sometimes paid in rice.

7.3.37 *"A Different Perception of Time"*

In pre-war living, some families had enough firewood stored for 10 years. Some people did not think much of walking two hours back and forth to do errands. In some regions, a household once every 20 years was in charge of producing charcoal for the community. Some people spent years making bedclothes out of royal fern cotton for their daughter's future marriage. The perception of time was different from the one we have today.

7.3.38 *"Gratitude and Reverence in Daily Life"*

In pre-war living, things were used with great care based on a feeling of gratitude and reverence. In the morning, fresh water from the well would first be brought to the local shrine as an offering, and parents would tell their children to pay respect to the gods and to Buddha. Today, we never lack for water and fewer people feel a sense of gratitude for the ability to use water freely. Gratitude and reverence are feelings that may have receded in people's everyday life.

7.3.39 *"A Different Notion of Luxury"*

In pre-war living, oil and sugar were luxuries, and rice cakes, salmon and sardins were seen as treats. Sweets such as dried persimmons were also regarded to be a luxury. Today, eating your stomach full, digging into high quality steak, eating fresh fish or shells, or eating foodstuff that is normally only available overseas are things that are regarded as luxury.

7.3.40 *"A Sense of Laxness and Generousity"*

In pre-war living, bedrooms would sometimes be fully visible from the street, and the fact that neighbors might disturb or bother each other was seen as a natural part of life in the community. A natural sense of laxness meant that people might sometimes dig up sweet potatoes in a neighbor's field, boil them and enjoy them without permission. Rarely did people in such cases claim ownership and launch one complaint after the other as might be the case today.

7.3.41   *"Moderation: Knowing Just How Much Is Enough"*

In pre-war living, gauging just how much would be enough, or what was the right amount, was part of daily life. For example, when enjoying sake, there was a way of drinking called "mokkiri", which hinted at the fact that if people stocked alcohol they would likely drink too much, so better just to drink as much as could be carried in one go, and thus, also, to enjoy just the right degree of tipsiness. Also, opening the door just so much as to adjust temperature to the right level, or obtaining just the right degree of warmth in the house by stirring up the embers in the open fireplace were ways in which people pursued "just the right degree or amount" in daily life.

7.3.42   *"A Clear Distinction Between the Extraordinary/ Celebratory (hare) and the Daily (ke)"*

In pre-war living, when celebrations such as the New Year or festivals took place, people might splurge, but on other days, life would be very simple and humble. The occasional luxuries of life were truly looked forward to.

7.4   What Is Necessary to Co-exist with Nature

These ways of living that are gradually being lost, can, further, be described in the following manner. There used to be ways of living in which people had a direct relationship with nature—they enjoyed nature, utilized nature and were prepared for natural disasters. These ways of living are about to vanish from modern society, in which people are ever further removed from nature.

Also, to have a workable relationship with nature, living and acting alone is not sufficient, and even a household is too small a unit. The size of a region or community, in which people join forces to associate with nature, is just the right scale. People in the past therefore took great care to set up joint mechanisms in the community to handle things indispensable in daily life, such as resources found in nature, or food and water, etc. These mechanisms for sharing and co-managing are also being lost.

Furthermore, living together with nature is in many ways a tough challenge. The conditions of nature change from day to day. People at times had to deal with the fury of nature and could not always expect enjoying a wholesome, fulfilling life.

Many unique ideas were needed to find joys and experience fulfillment in an everyday that might well be filled with repetitive work and seemingly dull days. This is the reason why people in the community together created annual events and gave each other distinct roles to enable a sense of fulfillment in everyday life. This was also what helped establish a strong community.

People's legs were the most common means of transportation, and since it was difficult to travel long distance, there was no choice but to live together with the nature found in and around your community. This is what gave birth to the wisdom of local production/local consumption, or to the effective, cyclical utilization of limited resources.

The instinctive desire to wish for your children and grandchildren to flourish, led people to trade with communities in other regions so as to allow for a mingling of blood, and to create organizations enabling young people to meet.

These are undoubtedly ways in which people mastered techniques of living in companionship with nature. Furthermore, people had devised mechanisms to pass on this knowledge to the next generation. Everyone was given a role, and children could learn all the required tasks while growing up. Consumers watching work in progress was a way of honing the skills of craftsmen.

Mentally, in order to satisfy people's urge to consume in a society with limited resources, clever ways of dealing with desires were conceived of, such as the distinction between *hare* (the celebratory) and *ke* (the daily), and the custom of showing gratitude and reverence toward nature, thus creating societal mechanisms to keep consumption at an acceptable level. Efforts were made to find just the right amount or know just how much was enough. A positive way of thinking which found joy even in the smallest things in life was essential to co-exist with nature.

The above elements were created in a long process of trial and error—probably there were many failures along the way—and came to constitute a sophisticated set of wisdom, techniques and values of living. This is, indeed, what is being lost today. And we in modern society have not even noticed this. Why? The reason is that while people in pre-war Japanese society faced constraints in all aspects of life and accumulated pieces of wisdom for living satisfactorily within those limits, we in modern society have built our civilization on the illusion that all constraints have been overcome.

7.5   Low Impact Technologies Found in Nature and Their Similarities with Pre-war Living

Let us point to an even more interesting fact. Knowing that pre-war living was a way of life in co-existence with nature, there is a possibility that we may find similarities to the patterns found in plants and other living things in nature.

The Nature Technology Research Consortium has created a database of low environmental impact technologies found in plants and animals or elsewhere in the natural world. In this database, data on some 300 types of technology were included

Resources/energy/collection and storage of information	Defenses and stability
Water collection, Light condensing, Use of light energy, Animal power, Heat insulating, Cold insulating, Water retaining, Detection of seasons, Place recognition, Wide view, Replenishment of enzymes, Mute effect, Data accumulation, Smell cancellation, Receiving signals	Stabilization, Control of insect pest, Suction, Adhesion, Resistance to external threats, Fire resistance, Wind resistance, Cold resistance, Impact resistance, Heat resistance, Light resistance, Resistance to micro-organisms/bacteria, Resistance to pull, Long term preservation, Blocking of light, Regulation of moisture, Insulation, Attack, Control of living tissue, Water repellency, Trapping, Covering/coating, Dormancy, Prevention of coagulation
Resources/energy/transmission and production of information	
Light emission, Electricity generation, Heat generation, Heat dispersal, Colour change, Odour, spraying, Generation of minerals, Generation of fibres, Generation of dynamic power, Transmission	**Shapes/organization/systems**
	Opening and closing, Expansion, Change of shape, Dispersal of function, Improvement of environment, Change of size, Change in quality, Use of gravity, Decomposition, Use of natural phenomena
Movement/disposal/circulation	
Hole digging, Movement, Floating, Prevention of reflux, Air conditioning, Regeneration, Reuse, Circulation, Removal, Sucking up, Convection, Self-healing, Recovery, Jumping, Separation, Filtering	**Efficiency (energy and resource conservation, lightness)**
	Efficient use of energy, Weight saving, Avoidance of air and water resistance, Efficient arrangement, Learning, High strength, Piercing, Self-sharpening

Fig. 7.2 Functional categories of nature technology

as of spring 2013 (http://nature-sr.com/). This data on technology, we will here call "nature technology".

In the 3.8 billion years since the birth of life on Earth, life has faced innumerable challenges in harsh environments, and it was through this process that living things developed the most efficient forms of technology and systems operating on limited resources and energy. That is, living things survived and thrived by adapting to the natural environment while developing the technologies required for the continuation of life. When we compare such nature technologies with the way of living experienced before the war, intriguing similarities can be discovered.

The technologies included in the above-mentioned database are classified into the functional categories found in Fig. 7.2. The 250 types of nature technology were first analyzed and broken into functional categories by specialists in the life sciences, in technology, and in environmental science and then integrated into the categories shown in the above chart.

Nature technology is divided into six main categories: Resources/energy/collection and storage of information, resources/energy/the transmission and production of information, movement/disposal/circulation, defenses and stability, shapes/organization/systems, and efficiency. Living things gather resources, energy and information from nature, store these, and use them when required for activity. When necessary, things that cannot be found in the appropriate form in nature are produced. In order to communicate with other living beings, information is also transmitted. These actions are conducted using technologies with low impact on the environment. Living things also try to move, and material that is no longer needed internally is rejected into the external environmental. Internally, resources are used cyclically. These processes are actually quite similar to what occurs in a human being; actions conducted to maintain life. Let us try now to compare these with pre-war living. With regard to "collection and storage", we find close similarities with elements of pre-war living such as "reading the signs of nature", "local production and consumption of food, fuel and timber", or "there are shed and storage houses". "Transmission of information and production" are quite similar to "the house is a place of production", "a useful garden", "songs and sounds as a part of life", or "small scale trading abounds". And, "movement/disposal/circulation" look similar to "circulating things", "a lot of time is spent walking", "there are mechanism and places encouraging people to meet", or "going back and forth between the city and rural villages".

Furthermore, living things defend themselves against external attacks and strive for stability. Human beings have also been pursuing the same kind of conduct. With regard to "defenses", we find similarities with elements of pre-war living such as "being prepared for natural disasters", "acting with half a year into the future in mind", or "several trades to live by and changes in everyday life". When we look at "stability", we find numerous concepts of pre-war living that correspond, such as "the pleasure of living in balance with nature's rhythm", "places that provide a sense of belonging for people in the community", "annual events are important", "companionship with other people as a source of vitality", "passing on knowledge to the next generation through daily life", "visitors are entertained at home", "a working relationship with fire", "gratitude and reverence in daily life", "a sense of laxness and generosity", or "moderation—knowing just how much is enough".

Naturally, the harsher the environment the more living things will pursue aspects of efficiency such as energy and resource conservation, lightweight structures, etc. These correspond to pre-war elements of living such as "wisdom to use nature in daily life", "sharing the mountains, fuel and water", "joint effort and mutual support", "living together with people who are not family is common", "maintenance", "no excess of things", "the shape of the house reflects the way of living", "delivery businesses", "things are sold by weight", or "value that is not counted in money".

And, finally, living things attempt to realize low environmental impact, and thus to survive and thrive, by adjusting shapes, structures and systems. These are similar to elements of pre-war living such as "proximity to other living things", "community is bonded by shared tasks and mutual support", "you create the community yourself", "everybody has a role", and "an entertaining street of shops".

There are also some elements of pre-war living that do not apply to the functional categories of nature technology—concepts such as "children have a world of their own", "different notions of luxury", or "clear distinctions between *hare* (the celebratory) and *ke* (the daily)". These are all concepts that relate to enjoyment in life.

When it comes to the technologies or methodologies required to enable long term survival in the environment, we find that human beings are not exceptional to plants and animals. In particular, looking at the ways of living in co-existence with nature found in pre-war Japan, we can perhaps say that these are quite similar to the low impact technologies and approaches taken by living things to enable long term survival and "thrival" in the natural environment. Pre-war living in Japan can be called a way of living shaped by nature.

In the previous chapter, we mentioned how people in modern society attach importance to "nature", "enjoyment", "a sense of belonging to society", and of these in particular to "nature" and "enjoyment", in their lifestyles. It appears that learning "enjoyment" from nature is difficult, but clearly people in their everyday lives have sought different forms of enjoyment in nature. A way of living in which nature and enjoyment are integrated is gradually coming into view, as is the role of each of these two aspects. This appears to be an approach in which people—as we have described previously—realize that they are being kept alive by nature, utilize nature effectively, and know how to deal with the challenges put forth by nature.

Bibliography

Furukawa R, Satou T (2012) 90 sai hiaringu no susume (An encouragement to interview 90-year olds). Nikkei Business Publications, Tokyo

Yuuki T (2008) Touhoku wo aruku (Walking the Tohoku Region), Shinjuku Shobo, Tokyo

Chapter 8
The Pursuit of Nature and Enjoyment: The Contours of a Wholesome, Fulfilling Lifestyle

Abstract The unconscious desire people have for nature and enjoyment was beautifully expressed in pre-war living, and we can thus say that pre-war lifestyles included such elements as nature, enjoyment and a sense of belonging to society. It has also become clear, though, that citizens of today would not find it easy to return to pre-war ways of living. The notion of the irreversibility of (perceptions of) quality of life makes it obvious that elements of life such as a sense of modernity and convenience are deemed to be insufficient. From the viewpoint of the values people espouse today, these two elements were not sufficiently present in pre-war living. While we may say that a positive future is best created from the nostalgia of a (supposedly) golden past, it is not easy to actually reshape the ways of living of old—with their unity with nature and built-in enjoyment—into future lifestyles. The role of lifestyle design is not to promote a return to the past, but to envision new lifestyles that incorporate updated versions of useful elements found in old ways of living. How, actually, to introduce such elements as, for example, enjoyment found in pre-war living into future lifestyles is a big task, but the interviews we conducted with people around the age of 90 (below, nonagenarians), and the results of our analysis of 2030 lifestyles provide a broad outline of the structure of such future lifestyles. Taking severe, environmental constraints as a point of departure, wholesome, fulfilling lifestyles are built on three pillars: (1) convenience, (2) a nature that has aesthetic and cultural value, and (3) the recognition that constraints help both yourself and others grow and urge you to nurture nature. Apparently, overcoming constraints gives birth to enjoyment and a sense of fulfillment in people's lives.

Keywords 2030 lifestyles • Fulfilling ways of living • Nature and enjoyment in everyday life • Pre-war lifestyles • The irreversibility of (perceptions of) quality of life • The mould of wholesome • The structure of lifestyles

E.H. Ishida and R. Furukawa, *Nature Technology: Creating a Fresh Approach to Technology and Lifestyle*, DOI 10.1007/978-4-431-54613-9_8, © Springer Japan 2013

8.1 Low Impact Pre-war Lifestyles

The ways of living we discovered through our interviews with nonagenarians, were indeed prime examples of lifestyles with low environmental impact. We can, therefore, assume that the structure of 2030 lifestyles may quite possibly resemble that of pre-war lifestyles. Using the same methodology as with 2030 lifestyles, we outlined pre-war lifestyles and, using a questionnaire survey (n = 1,000) of these lifestyles, analyzed the structure—or main characteristics—evident therein. In Fig. 8.1, the 40 constituting elements of lifestyles derived from our interviews with nonagenarians are plotted against both the 50 examples of 2030 lifestyles described in Chap. 6 and 40 examples of pre-war lifestyles, in each case showing (scoring) the degree of compatibility between constituting elements and lifestyles. Some major characteristics of pre-war lifestyles (the elements that scored lowest on the chart) include their tendency to require time and effort and to be seen as extremely inconvenient. Also, they are not seen as mainstream, modern styles of living, are unlikely to suit your personal preferences, and information is not abundant in society.

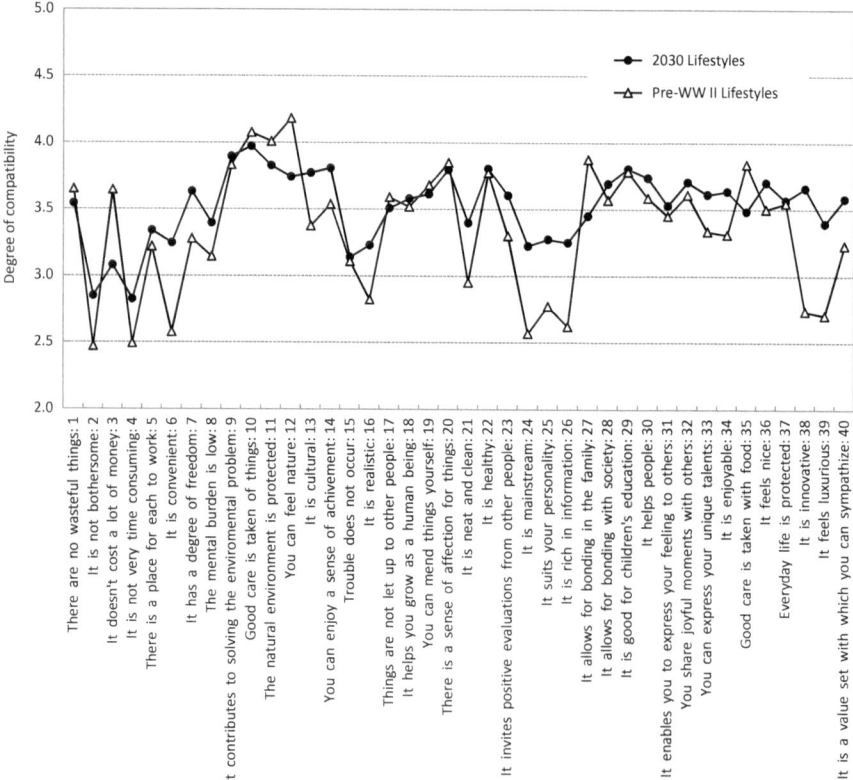

Fig. 8.1 The structure of pre-World War II lifestyles (40 examples) and 2030 lifestyles (50 examples)

Cluster number (Order of preference)	Factor 1			Factor 2			Factor 3			Factor 4			Social acceptability
Cluster 1	Naturo			Modernity/ mainstream			Enjoyment			Convenience			39.3
	10	9	11	24	16	38	34	36	40	2	4		
	4.01			2.99			3.59			2.73			
Cluster 2	Nature			Modernity/ mainstream			Social belonging			Convenience			33.1
	9	10	11	24	38	25	28	32	30	2	4		
	3.87			3.18			3.77			2.34			
Cluster 3	Enjoyment			Modernity/ mainstream			Nature			Convenience			29.3
	32	34	28	24	39	38	11	12	35	2	4	3	
	3.03			2.58			2.91			2.86			
Cluster 4	Social belonging			Modernity/ mainstream			Nature			Convenience			28.8
	32	29	27	24	38	39	11	12	9	4	2	3	
	3.38			2.75			3.6			3.04			
Cluster 5	Nature			Enjoyment			Convenience			Freedom			16.2
	10	11	12	34	36	40	4	2	6	7			
	3.85			2.99			2.4			3.01			

Fig. 8.2 Constituting factors of pre-World War II lifestyles and social acceptability evaluated with today's values

Furthermore, they are not seen as innovative and there is no sense of luxury. Perhaps this evaluation of pre-war living comes as no surprise. On the other hand, elements that scored highly included a sense of nature in people's lives, affection for things, health, strong bonds with family members, and the ability to share joyful moments with others. Also, such lifestyles are seen as good for children's upbringing. Furthermore, they are lifestyles in which care is taken not to waste food.

If the latent desires of today's citizens have not changed since pre-war days, these aspects should be detectable in pre-war living. Let us see if this holds true. Using cluster analysis, we categorized the 40 examples of pre-war lifestyles and, further, made a factor analysis for each cluster (Fig. 8.2). Looking at the factors which characterize these clusters and at the social acceptability patterns, we made the following discoveries.

The social acceptability of pre-war lifestyles is approximately 24 % lower than that of 2030 lifestyles. Looking at the factors that lower the social acceptability of pre-war lifestyles, it becomes clear that "it is modern and mainstream" and "convenience" are both evaluated negatively, to more or less the same degree. Among the lifestyles with lower social acceptability, we still find that where elements such as "nature", "enjoyment" and "a sense of belonging to society" are present, acceptance goes up. The same was the case when we analyzed the acceptance of 2030 lifestyles. Including only the element "nature" did not raise social acceptability.

From these results, we may say that since people in pre-war society also pursued "nature", "enjoyment" and "a sense of belonging to society", these were elements present in pre-war lifestyles. But, when people, based on present values, evaluate pre-war lifestyles, the fact that these do not feel modern/mainstream or convenient is likely the reason why social acceptability scores lower than for 2030 lifestyles.

8.2   The Difficulty of "Returning to the Past" and the Necessity of Backcasting

Some people are of the opinion that one possible solution to the environmental problem would be "returning to the past". The argument is that in order to reduce CO_2 emissions by 80 % and create a resource-cyclical society, we should just return to the way of living found in the Edo Period (1603–1868) or earlier. Looking at the data, however, we see that people with a modern value set would not readily accept pre-war lifestyles.

What will happen with social acceptability, though, if we use backcasting to design lifestyles, not aiming to return to the past, but in order to reshape elements of pre-war living into a modern version?

We compared three examples of lifestyles (one example of 2030 lifestyles and two pre-war) that emphasized "sharing important things in the community". The 2030 lifestyle was one in which the inhabitants had introduced shared batteries to use energy jointly in the community and support each other. We compared with two pre-war lifestyles that incorporated similar elements. One was the young people's mutual support organizations called *Yui*, which helped with agricultural tasks in the community while participants enjoyed the company of others; the other was the community collaboration taking place when thatched roofs were replaced or repaired. There was a large gap in social acceptability seen from the viewpoint of today's citizens: 70 % for the 2030 lifestyle, and for the two pre-war lifestyles, 12 % and 25 % respectively.

8.2.1   2030 Lifestyle <Shared Batteries> Social Acceptability 70 %

Sometime in the past, people started moving from large family houses to apartments to free themselves from various family obligations, and gradually the nuclear family was born. Life in traditional, long family houses was abandoned, and people started treasuring privacy. In the course of these developments, the bonds that existed in community vanished.

The installation of "community-shared batteries", which allowed people to protect their privacy while sharing energy infrastructure, is one development that has started putting a brake on this trend. In this system, solar panels are installed on each housing unit in the community, and the electricity generated is accumulated in storage batteries in each building. When a family, however, travels or for other reasons uses less energy than generated, the excess energy is automatically stored in common storage batteries. These storage batteries are of help to inhabitants when rainy days continue. Excess energy is not sold to the utility company, but, based on a Japanese spirit of generosity, energy that is left over belongs to all. Thus, a community which both protects privacy and enables mutual support is born.

8.2.2 Pre-war Lifestyle <Mutual Support Organizations> Social Acceptability 12 %

In the region where farmer M. lives, there are mutual support organizations called *Yui*, through which people in a neighborhood help each other when there is a need for many hands, such as at the time of rice planting etc. Helping each other is the natural way of things in the community. At the time of year when rice planting takes place, groups of about ten people share their breakfast in the field. Some 5–7 L of boiled rice and barley is prepared in a bowl, and *miso* soup is provided in a barrel with a lid.

There is also a young people's group in which the 15–20 year olds participate. M. is always looking forward to taking part in this group's activities. Some 30–40 people from M.'s village participate. Every year shortly after the New Year, the group holds an event in which people present their artistic skills or work, so when the farming season ends, the young people put all their energy into practicing songs or dances. Sometimes, young women and men who meet in this group end up marrying each other.

8.2.3 Pre-war Lifestyle <All the Village Helps with Roof Thatching> Social Acceptability 25 %

In the region where N. lives, almost all houses have thatched roofs and about every 20 years or so, the need arises for re-thatching. In between these major overhauls, smaller repairs works called *sashigaya* (straw insertion) must also be undertaken. This work is quite costly and requires a lot of time, so only about two houses in the community are fully re-thatched every year. To prepare for this, the whole village together cuts reeds at the end of November every year. When the actual thatching is done, about four skilled craftsmen come to help. An additional four people or so are needed for the work which takes between seven and ten days to complete. To cover the cost, all in the village contribute to a savings fund called *mujin* (mutual finance association).

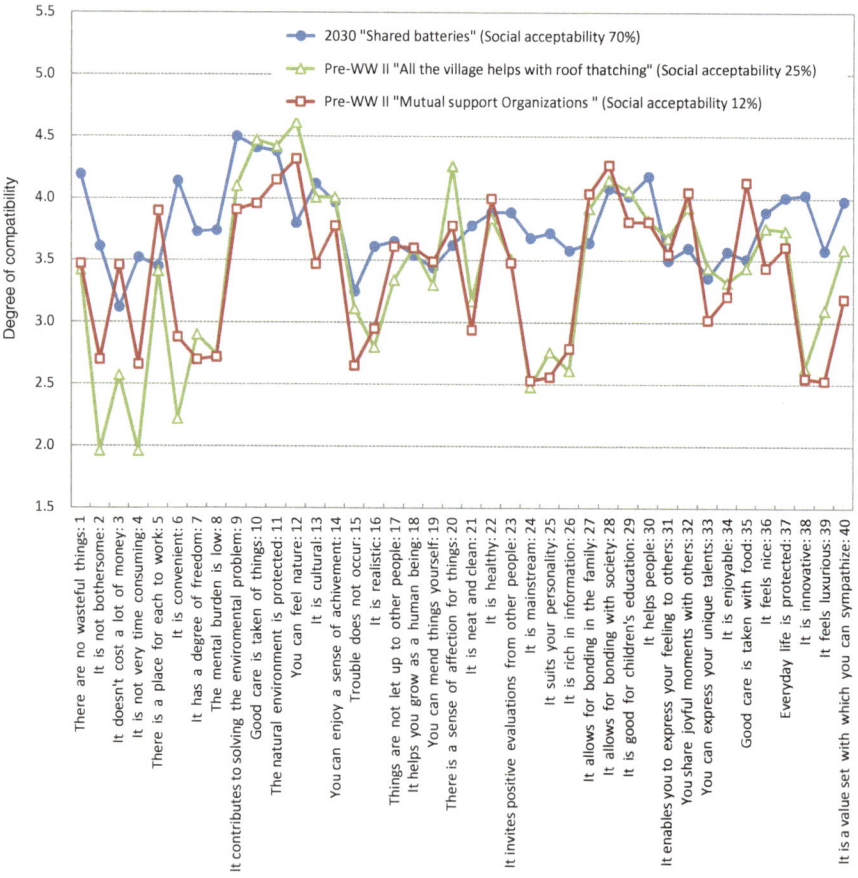

Fig. 8.3 The structure of lifestyles of sharing (comparison between 2030 and pre-World War II)

The ropes that hold the reeds are also made in the village from rice straw. The knitting of ropes cannot be done by children. You have to be about 15 or 16 before you become able to do this work.

Analyzing the structure of these lifestyles, it becomes clear where the significant differences are in the structures of 2030 lifestyles and pre-war lifestyles: Do everyday tasks take an extra effort and require time? Is living convenient or not? Is there a high degree of freedom? Is it mentally burdening? Does it feel modern and mainstream? Will you be able to adapt the lifestyle to suit your personal preferences? Is life information rich? Is there novelty and innovation? Is there a feeling of luxury? These are the evaluation items with large differences in scores. It is the low compatibility with—or applicability to—so many elements of pre-war life that cause big differences in social acceptability (Fig. 8.3).

Concepts such as mutual support or sharing are important factors in lowering environmental impact, but when these are merely seen as a return to past ways of living, the social acceptability scores drop drastically. On the other hand, we can also see from our findings that if a concept such as sharing is redesigned cleverly, it is possible to raise social acceptability scores significantly. That is, we need to take a fresh look at pre-war, low impact lifestyles, extract important concepts that enable low environmental impact, and then redesign these in order to raise social acceptability. This process is of the utmost importance—and it is here that lifestyle design comes to play a key role.

8.3   The Ability to Discover Enjoyment in Life

Summing up the results of our analysis of the social acceptability of 2030 and pre-war lifestyles, we can say that whether or not people desire a certain lifestyle depends on the degree to which elements such as "nature", "enjoyment", convenience", and "a feeling of modernity" are included. As one might have expected, people unconsciously seek these elements in their lives and will not readily accept that even one of these lacks to a significant degree.

At the same time, we must not forget that as the global environment degrades, the quality of enjoyment in modern society is also deteriorating. Our interviews with nonagenarians made it clear that people in modern society are gradually losing their ability to find enjoyment in daily life. The way in which most of the nonagenarians interviewed looked positively at life is fascinating. It is no exaggeration to say that people in their nineties are ingenious at discovering enjoyment. They have the power to find enjoyment in nature, things of daily life, time, and even hardship. Living in and with nature was mentally tough. To live wholesome, fulfilling lives under such harsh conditions, it appears that people mastered the ability to find enjoyment in even the smallest things in life. People in modern society, however, rarely find enjoyment in the chores and tasks of daily living and are getting used to having fun and joy ready-made and provided. Television, computer games etc., are prime examples of this trend. New games are released one after another, and people merely have to play one of these to have fun. As a result of this, is it not the case the case that our ability to discover enjoyment in daily life is weakening?

Let us take a closer look at the ways in which people in pre-war society found enjoyment in life. In the interviews with nonagenarians, we find statements such as: "In summer, we went barefoot to the fields or rice paddies. Stones were all over the place and in the rice paddies, there was a kind of straw that hurt when you stepped on it. We said "ooh and ouch", but we walked about without really being bothered too much by this. The back of our feet were energized by walking on dirt". The back of their feet hurt from walking barefoot in the rice paddy… it remains unclear whether the stones and straw stimulated certain acupuncture points under their feet, but the person interviewed expressed a positive way of thinking in saying that it hurt because their feet received energy from nature. We find an amazing ability to change a painful experience into enjoyment with positive thinking. To take another example: "When we went to plant rice in spring, we put a long bamboo stick in front of the nose of the horse—it was called a nose catcher. It was our job to control the horse using this stick and thus help with spring chores. This was very tough and great fun at the same time". Here, again, we find the ability to discover enjoyment in hardship.

People in pre-war society also had fun matching their wits against each other: "As kids we enjoyed hiding astringent persimmons in rolls of rice straw piled high and then ate them around the time they turned sweet. It was even more fun when you found and ate some that other kids had hidden. Sometimes, though, it was your own persimmons that were pilfered". The children enjoyed matching their wits against animals, too: "I think we had more interesting ways of playing than now, when people just watch TV. Our counterparts were animals and living things, and you had to trick them into getting caught. This meant we had to work our brains hard to succeed...I think you can say this was a true experience for us".

The excitement of observing things taking place in the community is another form of enjoyment we find in the interviews with nonagenarians: "Watching the *kamaboko* (boiled fish paste) being fried was also lots of fun". "There were craftsmen who were like repair smiths. They would go from house to house and fix pots that has become holey. As kids we used to enjoy watching how they brought their tools, heated metal with charcoal and poured it into moulds, and then used this to stop up the holes". "Walking home we would sing songs, and in winter we had great fun swinging our hand towels in the air until they froze and got stiff". In such ways, the children would enjoy watching the craftsmen at work in the shopping street or the different phenomena found in the surrounding nature. There was a greater abundance of nature than today for the children to watch and enjoy: "In the old days, every house had a small garden and persimmons and plums would ripen on trees there. Some gardens were covered with greenery almost like parks, and we had fun watching them. So, we really loved going for a walk". "There were so many small killifish in the rice paddies and we would scoop them up with our hand towels. It was great fun".

Furthermore, people in pre-war society enjoyed the gap between the daily routine and special occasions or treats. They enjoyed having sweets that were only eaten every now and then, the occasional luxury food, or clothes that were only worn on special days. The joy in this was amplified exactly because it was not food eaten or clothes worn every day. "In the equinoctial week, we enjoyed having sweet *ohagi* rice balls made for us and always dashed home from school". "We brought lunch boxes with us to elementary school. The side dishes (to go with the rice) were usually just salted *umeboshi* plums or some pickled vegetables, but every now and then there was fried egg in the box—this really excited us". "For the annual sports meeting only, the school prepared white skirts that we would wear while competing in the different events. It was the first time I wore a skirt so it was really enjoyable".

The people back then also had fun making a play spot out of any place at any time. "In the old days, we often heard children reading books aloud in the houses here and there in the village. And, on their way home from school, the children would sing verses they had learned at school in loud voices while walking. The roads, you see, were places for us to play". "It was about four kilometers to our school and we would all walk together for about one hour. We would say things like "if you lie, your tongue will be torn out" to our friends, and when we grew older

Fig. 8.4 Comparison of present and pre-World War II perceptions of enjoyment

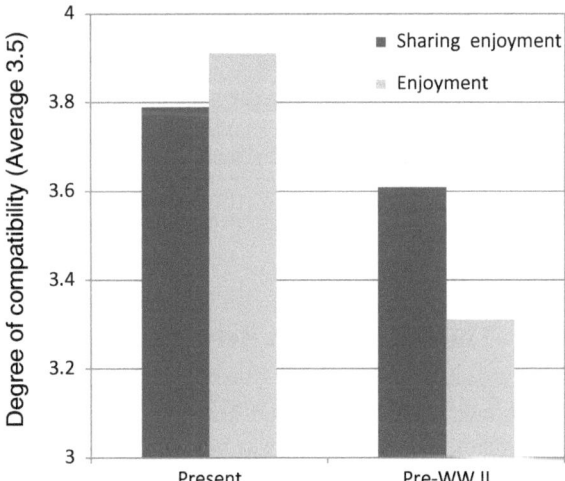

would teach the same to younger kids. We had quite a lot of fun on our way to and from school". "There were different Buddhist lectures—one was a Kannon (Goddess of Mercy) lecture held by a woman, another was by a person from the Bodhisattva sect, and others were like singing sermons (called *baikakou*). We also sometimes went traveling and enjoyed that".

Furthermore, the people of that time knew ways of sharing the joys of life. "Our father always told us to share your happiness with other people". In this way, wonderful pieces of wisdom were passed on.

Using the questionnaire survey we conducted on pre-war lifestyles (Fig. 8.4), we can compare with the degree of enjoyment of life and extent to which such joys are shared among people today.

The graph shows that there is a greater degree of (perceived) enjoyment today, whereas, in comparison, the ability to share enjoyment with others is slightly lower than in pre-war society. Evaluating pre-war lifestyles with the values people espouse today, the perception is that there was less enjoyment in life in the past, but that people more often shared enjoyment with others. Looking at the same issue from the vantage point of people living in pre-war society, we might find that there was actually more enjoyment in life than this data shows. If we, tentatively, accept a comparison from the viewpoint of modern values, we still find that in pre-war society, emphasis was placed on sharing enjoyment rather than enjoyment itself.

Among the nonagenarians, many people said "we always looked forward to the festivals", and these were indeed events that incorporated many such elements of enjoyment. The words of one interviewee remain vivid in our memory and describe how these old people are supremely confident of their ability to discover enjoyment in all aspects of life: "When I was a child, I did just what I wanted to, but also now I am enjoying every day I live. The world is such a fascinating place. I intend to live until I am a hundred years old".

People who lived in pre-war Japan were, indeed, ingenious at finding enjoyment even when facing constraints (Fig. 8.5).

> - Changing hardship to enjoyment through positive thinking
> - Having fun matching your wits against others
> - Enjoying observation
> - Enjoying the gaps present in life and creating gaps to enjoy
> - Creating places to play out of everyday scenes, everywhere and anytime
> - Sharing enjoyment with others

Fig. 8.5 Ways of discovering enjoyment in life

8.4   The Relationship Between Nature and Enjoyment

When considering where and how people in low impact, pre-war society found enjoyment, we see that this was not only through interaction with other people, but also from nature itself. In nature, there are numerous living things. People used to have fun matching their wits against other living things, and even enjoyed it when they could not have their way in relation to animals or other living things. People found enjoyment in nature as an object of observation. The natural environment was harsh, both the quantity and quality of harvests fluctuated, and people suffered a lot from this. However, there would always be times of bountiful harvests and a season of abundance would arrive. In this way, gaps between times of hardship and times of plenty inevitably arose. People were able to transform these gaps into enjoyment. Nature was an easy place in which to discover enjoyment—it was, indeed, a reservoir of joys. One might even say that it was the very harsh conditions found in nature that gave birth to people's ability to discover enjoyment in daily life. Possibly, it was, indeed, nature which shaped the form of enjoyment in society.

When we work on lifestyle design, if joyful moments in life are merely designed to be provided to citizens, we will end up in the same situation as today. Unless we design lifestyles that cultivate people's ability to discover enjoyment in everyday life, we may not be able to shift present lifestyles in a positive direction.

The way of living in pre-war society, and the enjoyment found therein, are recorded in written works. Some examples are "Recollections by Kiyokata Kaburagi" by the painter of *Ukiyoe* and Japanese paintings, Kiyokata Kaburagi (1878–1972), or "Letters to a Person Behind Bars" by novelist and commentator Yuriko Miyamoto (1899–1951). Kiyokata Kaburagi was a painter of traditional Japanese paintings who through his entire career depicted the customs, daily rituals and ways of life in old downtown Tokyo as well as modern-looking beauties, but he also left a collection of essays on ways of life and enjoyment in old downtown Tokyo that were gradually disappearing. In "Recollections of Kiyokata Kaburagi", the relationship between nature and enjoyment in early Showa Tokyo (The Showa Era started in 1926) comes into view. From this collection of essays, we extracted all examples of enjoyment and similar phrases and found that numerous of these were closely related to nature. This points to the fact that ways to enjoy nature or seasonal changes were not only pursued in a rural prefecture such as Miyagi, but also in a metropolis like Tokyo. The ways in which people in Kaburagi's writings found enjoyment in nature or in seasonal changes are, largely, the same as those described in Fig. 8.5. Let us take a look at a few examples.

8.4.1   From "Recollections of Kiyokata Kaburagi"

Back in the days when the cold didn't get to me as much, winter was not as unpleasant as it is nowadays. First or all, snow—which I loved—would fall. Sometimes, it would not be snow that was formed, but ice-cold droplets of hail that would stick to the maple branches like a frozen necklace conveying the rustling sound of the north wind. Looking at this scenery through my glass door was one of the pleasures of the winter season. (January, Showa 16 (1941).

Even just looking out over the city from the 1st floor of my house, I am pleased to see that Tokyo is still, also, a green world. Gingko trees as tall as giants can be seen, here and there, soaring towards the sky. In the sky south of my window, as well, my neighbor's huge gingko tree stands lushly green and vigorous in figure. In the garden of the house where I live there are no trees or plants that particularly fascinate me, but in my neighbor's garden—although no special care has been taken—there are gingko trees, maples, mountain camellia, plum trees, and oaks; all of them steadily adding on age. I also see bamboo grass and jetbead— and all of them are calmly content with their life in that place. The owner of the garden is unaware of all this, but the greenery gives me immense joy. (May, Showa 2 (1927).

In the evening bath, I wash away the fatigue of a day's work and let it flow along with my sweat into the sewer. I sling a *yukata* kimono over my shoulders and enjoy the cool of night while watching the stars. This is, indeed, very much like a simple, nocturnal social gathering. (July, Showa 4 (1929).

When the days gradually turn colder, a longing for the warmth of the fire is born in the heart of all and everyone. In the countryside, people will sit at the hearth-side, in the city in front of their stoves—they may sit around the long charcoal *hibachi* hearth, or outdoors sur- rounding the open-air fire, and the brightly burning warm fire will serve as a go-between and connect people whose thoughts and feelings are normally scattered. Surrounding the fire, people will commune with each other, and the intimacy between one human and another that our ancestors have felt again and again since the origin of time is, here, received from the past and passed on to the future. (October, Showa 12 (1937).

These writings express ways of discovering joys with low environmental impact that appear to be fading today. To what extent do we today know how to find such low impact joys in life? Finding enjoyment in nature is one of the most important aspects of life that we must not lose.

8.5   The Contours of Wholesome, Fulfilling Living

In the above, we have seen how there is a certain continuity, or correlation, between the values people today unconsciously seek (derived from our 2030 lifestyle research)— "convenience", "nature", "enjoyment", "social belonging", "self-growth"—and the ways of pre-war living found in our interviews with nonagenarians. The 70 keywords we extracted from the interviews can, largely, be divided into categories such as "relations to nature", "bonds with the community and family", "ways of living", "ways of working", "the relationship between life and death", and as we visited numerous evacuation shelters after the Great East Japan Earthquake, we found that in shelters were the mood was positive and people were smiling despite having lost everything, these keywords were vividly present. This experience convinced us that these are keywords of living that we Japanese must not lose. (At the moment, interviews with nonagenarians are also being conducted outside of Japan, and we are very much looking forward to seeing what keywords will be extracted from these interviews, and whether or not they differ from the ones found in Japan). The underlying current of all these keywords can be summarized as a way of living in which people: "know that they are given life by nature, utilize nature cleverly in daily life, and enjoy dealing with the challenges put forth by nature".

At the same time, however, it is a fact that people today will not readily accept pre-war ways of living in their original form. Here we find a large barrier caused by the irreversibility of (perceptions of) quality of life. More specifically, a pre-war lifestyle does not feel modern and mainstream to people today, it takes effort and time, and novelty or a feeling of luxury in everyday life are not found. It is also true, though, that when people today look at pre-war lifestyles, they find positive elements such as an abundance of nature in everyday life, an affection for things, healthy living, bonds within the family, a sharing of the joys in life, and good care taken with food and of children. Apart from "convenience", all the desires that we, through our analysis of 2030 lifestyles, found people unconsciously have today are, thus, actually seen to have been present in pre-war living.

If we, then, try to superimpose these two lifestyles, which share some common traits, on each other, what kind of measure or scale will this give us to evaluate "wholesome, fulfilling ways of living"?

For example, the gap that becomes evident when people today say that pre-war lifestyles, in their unadulterated form, do not appear to be modern or mainstream or that there is "no feeling of novelty or luxury in life", should come as no surprise, and there is little need to discuss this in detail here (There is a need, though, to translate valuable elements found in pre-war lifestyles into present day values and contemplate the issue anew). The same is true for aspects such as "it requires effort and time" or "it is inconvenient", but, on the other hand, we also found that there are aspects of pre-war living closely correlating to the latent desires found when contemplating 2030 lifestyles, such as the ways in which enjoyment is created and shared in everyday life, and not merely provided mindlessly. In the same way, it is evident that when people strongly seek nature in their lives, it is not sufficient to think of this as a desire to be in contact with nature (knowing that nature is what gives you life, or dealing with the challenges put forth by nature). There is clearly a need for an approach which utilizes nature cleverly; that is, in which people, facing various constraints, have the wisdom to fully harness the powers of nature.

Contemplating these issues again and again, the contours of a wholesome, fulfilling way of life—though still tentative—have started to emerge (Fig. 8.6). According to

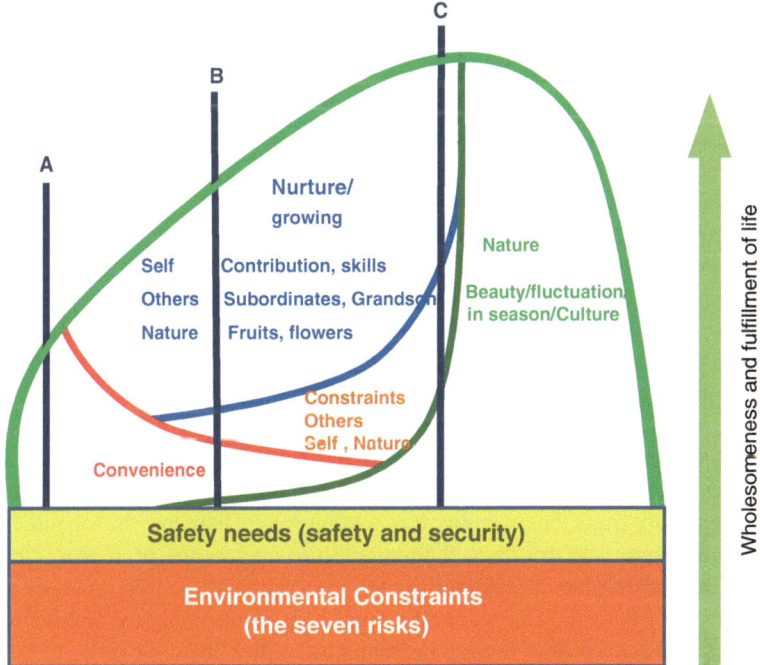

Fig. 8.6   The contours of wholesome, fulfilling living

recent research in positive psychology, the happiness experienced by people is 50 % due to genetic factors, 10 % comes from social status, fame or income, and some 40 % from behavioral changes. Figure 8.6 depicts issues relating to exactly these last 40 %.

Naturally, the contours of a wholesome, fulfilling way of living described here take the severe environmental constraints of 2030 as a point of departure. Also, when we talk of a wholesome, fulfilling way of living, we are obviously not discussing issues at the borderline between life and death. Therefore, thinking of Maslow's hierarchy of needs, our discussion presupposes that at least both physiological and safety needs (the two lowest rungs on Maslow's pyramid) are met. Thus, we are able to establish the constituting elements of a wholesome, fulfilling way of living based on the acknowledgment of global environmental constraints and the assumption that safety needs are met. The first two constituting element are "convenience" and "nature". Both in pre-war and in 2030 lifestyles, people seek convenience and nature in their everyday lives. The nature we are talking about here, though, is a nature which possesses both cultural and aesthetic value—that is, people derive mental satisfaction from the beauty of nature and are soothed by nature. The third constituting element is "nurture/growth". It indicates that people find enjoyment in and experience a sense of fulfillment or achievement from the nurture of themselves and others, and of nature. In our analysis of 2030 lifestyles, we found that people, unconsciously, seek enjoyment in their lives, but it is clear that what is meant by this is, of course, not just the fun of electronic games or movies or the internet. It is the kind of enjoyment—evident from our pre-war lifestyle analysis—that people create through participation and engagement. This is, indeed, what is meant when we here say "nurture" or "growth". There is, however, one condition that has to be met when we here talk about "nurture/growth". Electronic games may be fun, but people say that they feel hollow when playing for a long time. Why? Is it perhaps because people do not gain a sense of growing or developing themselves when playing such games? People do use their fingers and such games may be good to prevent early ageing, but we believe the sense of achievement or fulfillment derived is scarce. What is it that allows for a richer experience? We believe it is the existence of constraints. Constraints are needed as part of the element "nurture/growth"—and here we are talking about self-constraints, constraints arising from your relationship to others, and constraints caused by nature.

Let us take the example of growing a tree bearing delicious fruits—nurture/growth of nature. It would be wonderful if people only needed to water the tree to make it bear tasty fruits, but in reality it is not that easy. There may be too much rain, or too many sunny and dry days, the temperature may be too high or too low for the tree, and the volume of fertilizer needed also varies as the tree grows or due to the condition of the soil. These are all natural constraints. Only by mobilizing our wisdom may we overcome these constraints, or utilize them cleverly, to become able to harvest plentiful, delicious fruit. It is as a result of this process that various feelings of fulfillment or achievement are born.

Or, we set up constraints for ourselves in our work. We set stretching goals slightly above our present capability and gain a sense of fulfillment and achievement from this process of self-growth. In the community, people may try to sort out the many different opinions of people, and establish new rules or initiate new activities from this effort, thus nurturing others in the process, strengthening bonds, and even growing themselves. Using your wisdom or skills to gain a sense of achievement from situations that are slightly constrained or inconvenient is, we believe, exactly what leads to the kind of enjoyment people unconsciously seek.

The degree of wholesomeness or fulfillment in life is depicted on the vertical axis of Fig. 8.6, and thus moving from points A through B to C indicates an ever higher degree of wholesomeness and fulfillment. A, for example, may represent an eco-logical washing machine. This is a product where the conventional functions of a washing machine have been made more ecologically sound. Although this product does take as its point of departure the global environmental constraints, there are no changes in the basic washing machine functions—it is a technology in which some parts have merely been substituted for others; a product that merely pursues convenience. As such, it may very well end up becoming a product that triggers eco-dilemmas. What kind of technologies do we then find in the area B? If we start from the assumption that a product here meets requirements created by global environ-mental constraints, we can, for example, compare a quartz watch with a hand-wound watch. In most cases, we will find a high degree of precision in the quartz watch and the pursuit of craftsmanship in the hand-wound watch...we are thus looking at two different sets of measurement. Looking at the purpose of the watch—to keep track of the flow of time—we should, however, be able to judge both watches with the same set of measurements. How can we, then, evaluate the difference between the two in this diagram? If we look at a normal use situation, the enhanced accuracy of the quartz watch will lead to an increase in convenience, and in this case, the watch should be placed in area A just like the ecological washing machine. With regards to the hand-wound watch, there is a small constraint in the sense that we have to rewind the watch every day, but looking at the watch every day when rewinding, we start getting somewhat intrigued by the way in which it may be a little ahead or a little behind. Every day, as we rewind the watch, it provides us with a chance to think of the day's schedule. Thus, thanks to the existence of a constraint or inconvenience, we start feeling affection for the watch. This is also a form of what we call "nurture/growth". If, however, the constraint becomes so significant that we can no longer control it, it simply becomes an annoyance. Here the degree or extent of the constraint becomes extremely important. Think, for example, of the "thousand day-walk" from the famous temple Hieizan Enryakuji, which is said to be the toughest Buddhist ascetic practice lasting a total of 7 years to complete. It is said that a person who discontinues the practice before completion will die, whereas the completion of this harsh practice will supposedly lead to the reincarnation of the individual as the figure Ajari, who in Buddhism is regarded to be the embodiment of the Acalathe God of Fire (also known as *Fudomyoo*, a Buddhist guardian deity). During this religious practice, the apprentice has to endure a total of nine days with no food, no water, no sleep, and no lying down, and in the 7 years walks a distance equivalent to the circumference of the Earth. This could be an experience that is positioned in area C in the diagram—with your own willpower you overcome immense constraints and reach the very territory of the divine. Needless to say, however, the constraints faced here are so tough that they would, for most people in society, represent nothing but utter suffering.

From this angle, we may perhaps say that the first step in creating wholesome, fulfilling ways of living is to cease the excessive "outsourcing" of activities and skills in society. In order to gain comfort and convenience, we can use money to acquire the desired technologies or services. Needless to say, the resource efficiency of this outsourcing is worse than if each of us were to make our own efforts and if it progresses will lead to an exponential growth in the consumption of resources and energy. Using money as a tool, we outsource all and everything, and today appear to believe that we can even buy time with money. Although we clearly do not, actually, welcome this situation, reality is that we are running at full speed in the opposite direction of our unconscious desires. Why is this happening? The reason is that technology and services are being provided which appear to be arguing that constraints are an evil. Removing constraints is the very essence of mainstream business today. With such an approach, it becomes increasingly difficult for people to enjoy wholesome, fulfilling lives. Needless to say, businesses or services that help people deal with constraints that cannot be overcome may have a raison d'etre in society, but the important thing is that in order to create wholesome, fulfilling ways of living, constraints are unavoidable; they are indeed required in order for us to make the shift from a lifestyle of dependency (on the technologies and companies to which we have outsourced our competencies) to one of independence. Needed today are services and technologies that help enable this transition.

At the same time, we citizens ourselves need to make a personal effort to make the transition to an independent lifestyle. We are not arguing that all forms of outsourcing should be discarded. But we do believe an effort must be made to consider how, little by little, we may regain some of what we have lost through outsourcing. From time to time, we should enjoy cooking with a proper, home-made soup base, relish food in season grown in a small kitchen garden on the veranda, try how it feels to sharpen a pencil with a knife, cook a fish ourselves, write a letter with a fountain pen, or try to make a book shelf or chair ourselves...Creating such "constraints" and using them to enjoy life is, indeed, what gives birth to a greater degree of wholesomeness and fulfillment in our lives.

The only way we can become able to survive on this Earth is to enter and take part in the cycles of nature. This challenge depends on the degree to which we are able to reduce outsourcing, and on whether we can find other measurements in life than money.

The diagram attempting to outline the contours of wholesome, fulfilling living is still very incomplete. It is our goal to use this approach to evaluate various products on the market today. We believe this would also be useful as a tool to evaluate the technologies developed on the basis of lifestyles outlined using the backcasting methodology.

If, when using this tool, one were to find that the element "nurture/growth", for instance, was insufficient, and thus would like to "add" a bit of this to a technology or product, this would not be easy, though. The diagram is fundamentally a tool to evaluate completed or existing technologies, services etc., and is not a tool for developing products or services. For such development, a lifestyle design approach using the backcasting methodology is required.

Bibliography

Miyamoto Y (2012) Miyamoto Yuriko Zenshuu Dai 21 Kan (Complete works of Yuriko Miyamoto, vol 21), Nikkei Business Publishing, Tokyo

Yamada H (ed) (1987) Kaburagi Kiyokata Zuihitsushuu (Recollections of Kiyokata Kaburagi). Iwanami Shoten, Tokyo

Yuki T (2008) Tohoku wo Aruku (Walking Tohoku). Shinjuku Shobo, Tokyo

Chapter 9
The Transition to New Lifestyles: Transitional Technology

Abstract While global environmental constraints become ever more severe, lifestyles continue to be dominated by a forecasting mentality making a shift in direction difficult. As a result of this, all human activity, including corporate activity, tends towards partial solutions, and since we are not aiming at creating total system solutions, it will be difficult, as the severity of environmental constraints intensifies further, to realize the lifestyles that people deep down desire. While many changes are taking place as a direct result of emerging environmental constraints, the kind of world people truly desire will not be created simply from constraints, and in a worst case scenario, we may, on our own initiative, move further down the path of civilizational collapse.

Today, we cannot just wait for the weeding out of unsustainable technology and lifestyles but have to generate this weeding out intentionally and contemplate ways in which to realize attractive lifestyles even under environmental constraints. That is, we need innovation that shifts lifestyles in the direction of low environmental impact. The technology required to enable this must be sought in nature with its perfectly cyclical loops driven with a minimum input of energy, and considering the way in which most people unconsciously yearn for nature in their lives, this also appears to be the ultimate shape of technology. There are, however, significant hurdles when we try to move urgently in this direction. In order to make this shift, innovation is needed that can create lifestyles with total system solutions based on backcasting, using existing technologies. We call technology which enables the completion of this intermediate step on our way to sustainability "transitional technology". Some technologies of this kind have already seen the light of day. These are, for example, technologies using weak energies, technologies applying a community approach as learned from our interviews with nonagenarians, or a combination of these two—the smart city.

Keywords Lifestyles • Parklets • Partial solutions • Selection (weeding out, creative destruction) • Smart city • Total system solutions • Transitional technology • Weak energies

E.H. Ishida and R. Furukawa, *Nature Technology: Creating a Fresh Approach to Technology and Lifestyle*, DOI 10.1007/978-4-431-54613-9_9, © Springer Japan 2013

9.1 Environmental Constraints and Creative Destruction

Environmental constraints do not merely influence products and services, but impact the entire way of living in society. When, for example, the cost of energy rises, people begin considering which electric appliances to use less and which to continue to use as hitherto. Thus, as environmental constraints intensify, this will influence not only individual appliances but entire lifestyles. As the Great East Japan Earthquake suddenly occurred on March 11th, 2011, people in the disaster area were forced, in an instant, to live under severe constraints, but, generally speaking, environmental constraints are intensifying gradually over years. Under this influence, corporations are starting to change. In many corporations departments such as environmental or CSR divisions have been established and many have started moving in the direction of environmentally considerate conduct. Many ecological products, such as energy or resource conserving products, have started appearing. It is rather meaningless to try to predict the exact year by when environmental constraints will manifest themselves, but we believe that humanity will probably be impacted by severe constraints around the year 2030. Thus, if present trends continue and constraints gradually intensify as we move toward 2030, the number of ecological products will increase little by little. It will, obviously, be too late if proper countermeasures are only initiated in a gradual fashion as constraints emerge in the years ahead. Taking the emissions of greenhouse gases, and related constraints, as just one example, unless Japan reduces emissions by more than 50 % by 2030, the country will not be able to contribute to the alleviation of global warming. Innovation also requires time. Furthermore, additional time is needed to innovate lifestyles.

Let us try to imagine what would happen if environmental constraints gradually emerged. As described in Fig. 9.1, products in a particular category that are not energy or resource conserving would, first of all, no longer sell in the market. This is a phase in which, to take one example, ecological air conditioners would sell better than conventional ones. That is, the weeding out of technologies that are not optimized would take place. As environmental constraints get more severe, the weeding out of products across product categories would occur. This is a phase in which consumers would start comparing, for example, an air conditioner with a refrigerator and consider which one to choose and which one to stop using. When the constraints get even more severe, it is likely that a weeding out that includes the innovation of lifestyles would take place. That is, a phase in which a shift would occur from an energy consuming lifestyle to one that does not use as much energy.

As mentioned above, lifestyles are dominated by a forecasting mentality, and a shift in direction is therefore not easily achieved. Also, human activity, including corporate activity, is tending towards partial solutions, and since we are not seeing a shift toward total system solutions, the products that people actually desire may be weeded out as environmental constraints continue to intensify, and there is a significant possibility that, ultimately, the lifestyles people yearn for cannot be realized.

Innovation influenced by the emergence of environmental constraints is progressing, but the kind of world people truly desire will not arrive in this way, and in a worst

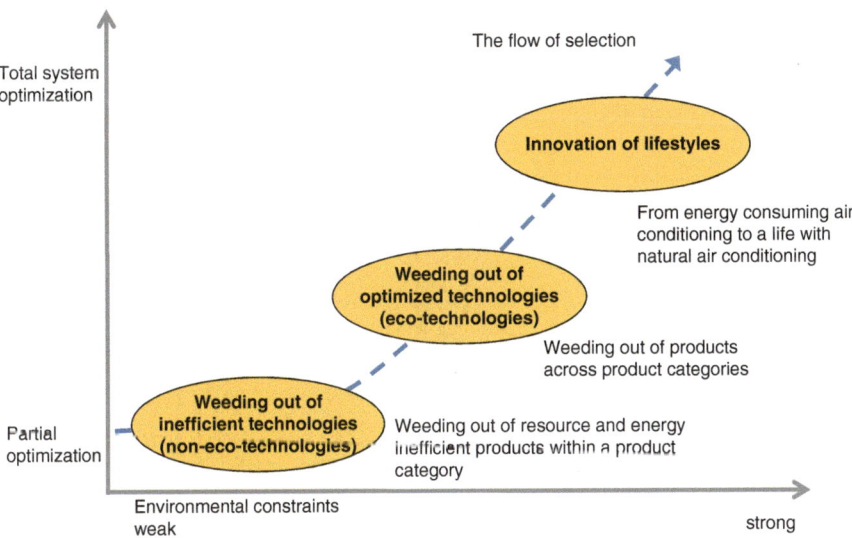

Fig. 9.1   Selection process in the market under environmental constraints

case scenario, we may, on our own initiative, move further down the path of civilizational collapse.

Therefore, we cannot wait for the weeding out—the necessary creative destruction—to happen by itself, we must intentionally create a process of selection and contemplate ways of realizing desired lifestyles even in the face of environmental constraints. That is, we need innovation that changes lifestyles in the direction of low environmental impact (Fig. 9.1).

9.2   Transitional Technology

What kind of innovation would be able to change lifestyles towards low environmental impact? It involves products and services created with technologies that enable a shift from today's lifestyles to "lifestyle X" under future environmental constraints (Fig. 9.2). In much the same way as the Sony Walkman made people take their music with them outdoors—thus causing an innovation in lifestyle—what we need today is the development of technologies that, through the introduction of a new product or service, enable a shift of lifestyle towards low environmental impact.

Ideally, such technology would take nature, with its perfect cycles driven with a minimum of environmental impact, as its model, but making this kind of transition in one go is rather too ambitious. We need technology that can help make the transition smooth—and this is what we call transitional technology. This is, as we shall see in detail later, existing technology which—through for example the utilization of weak

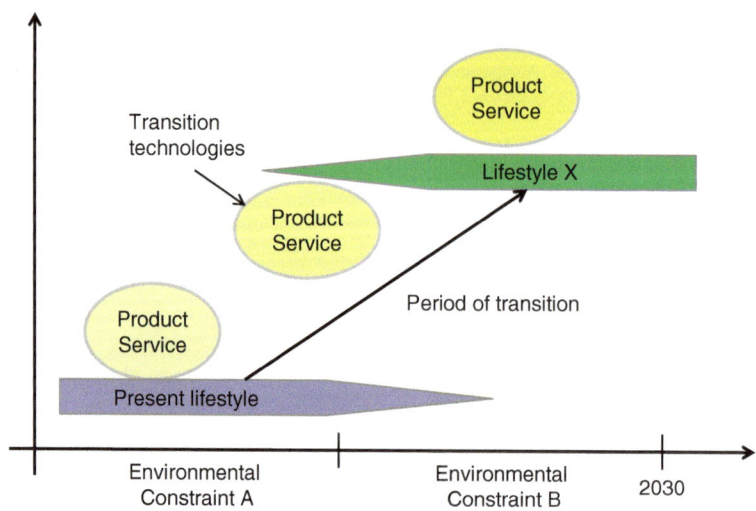

Fig. 9.2 The role of transitional technology

energy, is still able to contribute to lifestyle innovation. Unfortunately, only few transitional technologies have so far seen the light of day. There are many technologies making a certain product itself more ecological, but we can safely say that there are almost none that help shift entire lifestyles towards low environmental impact. The reason is, of course, that the lifestyles towards which innovation is needed have not been envisioned and, thus, remain out of sight.

9.3   The Present Status of Environmental Innovation

Here, an understanding of the present trends in environmental innovation becomes necessary. Global environmental problems are issues concerning the entire world. Since Japan is dependent on the import of energy and food, it is inevitable that the country will be influenced by severe environmental constraints if the gravity of global environmental problems progresses. Already, environmental constraints are causing people's awareness to change. This change is characterized by the emergence of "environmental /green needs" through which people wish to contribute to the reduction of environmental impact. Despite the fact that people still have a somewhat distorted perception of environmental problems, environmental/green needs are nevertheless gradually starting to influence innovation,. This innovation is starting to shift from a situation in which the focus was on the optimization of individual actions of consumers to a focus on the optimization of the behavior of groups of consumers, sometimes involving entire cities or regions. For example, a shift is taking place from devices with high energy consumption to devices incorporating technologies enabling lower environmental impact, including the usage patterns of

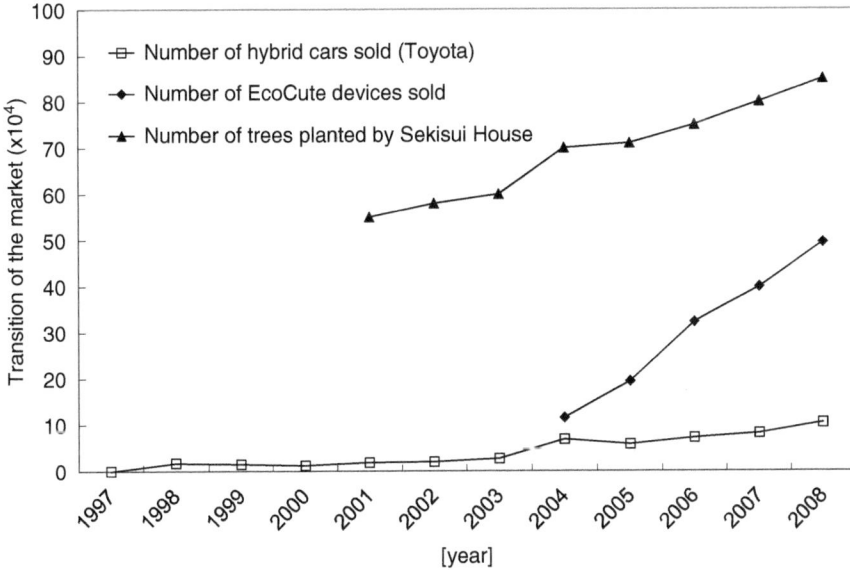

Fig. 9.3 Sales trends of three environmentally considerate products (Hybrid cars, Plant-five-trees, EcoCute) (Furukawa 2010)

the device. However, it is only a tiny part of products in which this kind of total system optimization of an entire device including its usage is taking place. In most cases, the innovation focuses on partial optimization aiming to lower the environmental impact of individual products, and as a result thereof ends up causing eco-dilemmas. The general trend is not towards total system optimization aiming at the lowering of the environmental impact of entire lifestyles. If we continue with innovation which does not take constraints into account and aims only to satisfy the present desires of consumers and conventional development ideas of engineers, we will not be able to cease the excessive expansion of human activity. This does not mean that innovation as such should be discontinued—what we need to consider is how to generate innovation that makes it possible to reduce the environmental impact of lifestyles. The important task is how to make the transition to an innovation system which helps realize wholesome, fulfilling ways of living even under environmental constraints.

It is difficult to pin down exactly when environmental/green needs emerged and started expanding, but in Japan, data suggests that the build-up of such needs gained speed around 2004 or 2005. One form of corporate environmental action has been the sales of environmentally considerate products. In Fig. 9.3, we have plotted the sales trends of Toyota hybrid cars (including also other brands than the Prius), of a service called "Plant-five-trees" by house manufacturer Sekisui House, and of EcoCute, an energy efficient heat pump (Fig. 9.3).

The Toyota hybrid car, Prius, was first put on sale in 1997. In 2003, the second generation Prius was launched, and in 2004 sales started to increase. After that, the use of hybrid technology was expanded to a hybrid minivan, a sports utility vehicle and a sedan, and in 2009, the third generation Prius was released. In August 2009, the cumulative domestic sales of hybrid cars in Japan reached two million units. According to Toyota's "Sustainability Report 2009", the company sold a total of 1.8 million hybrid cars between 1997 and April 2009 and through this contributed to a reduction of CO_2 of more than 10 million tons (Toyota calculations).

Sekisui House in 2001 initiated a customer-oriented program called "*Gohon-no-ki*" (Plant-five-trees), which—with small, Japanese hilly woodlands (*satoyama*) as a model—planted indigenous or naturally occurring trees suited to the local environment. The marketing slogan was "Three trees for the birds, two for the butterflies—plant indigenous trees that suit the environment in which you live", and the program was a proposal encouraging customers to help revive the natural environment. As a result of this program, the number of trees planted by Sekisui House grew year by year, and the company became the largest corporate planter of trees in Japan. According to interview surveys conducted by Sekisui House, about 50 % of the trees planted annually, as indicated in Fig. 9.3, were planted through the "Plant-five-trees" program. Some of the customers, who planted five trees as part of the program, now use it for nature observation with children. With this program it is not possible, as it was with the Prius, to show the exact quantitative contribution to CO_2 reductions, but it is a first step in the direction of a symbiosis with nature.

EcoCute is the generic term for a heat pump which uses heat extracted from the air to heat water, and which, as a coolant, uses CO_2 instead of freon gas. The number of these heat pumps sold for household use in Japan was 120,000 in 2004, but by 2008 had risen to 500,000 units per year.

These are but three examples, but there are numerous others that support the claim that the number of products responding to environmental/green needs have increased in the market recently.

Also, regarding the awareness of citizens, data from opinion surveys substantiate the same claim. In the annual "Survey on Social Awareness" conducted by the Japanese Cabinet Office (Here, data from 2009), when asked whether or not respondents "would like to contribute to society through their actions", 69.3 % responded "yes". When these 69.3 % (4,080 respondents) were then asked in what way they would like to contribute to society, an increasing number of people since 2006 responded "through activities related to nature or environmental protection", the percentage of which rose to 41.6 % in 2009.

Interview surveys conducted by this book's authors indicate that the environmental/green needs of consumers in relation to energy intensive products (air conditioners, refrigerators, TV-sets) from the latter half of the 1990s started to actually reach engineers working on environmentally considerate products—among other channels through the salespeople working in large electronics retails stores (Furukawa et al. 2008). As a result of such trends, corporations immediately started shifting the direction of innovation. For energy intensive devices such as air conditioners, refrigerators, TV-sets and lighting, an environmental regulation called the "Top-runner Standard" was introduced, but even under heavy regulatory influence, corporations did not cease innovating merely because they had met this standard. Through the strong influence of environmental/green needs, the nature of innovation for energy conservation changed. And, recently, the innovation of energy intensive products has shifted from a focus on technological development to improve energy efficiency of the appliance itself, to one in which the focus is on the reduction of energy loss in the use phase (Ito 2009). Mitsubishi Electric's air conditioner, Move-Eye, for example, is equipped with a sensor enabling air conditioning which targets the area in a room where people actually are. Earlier, the technology installed made it possible only to cool the entire room, but a shift has been made to air conditioning which follows the movement of the user, thus reducing energy waste. This is not a case in which environmental regulation has had a direct influence; rather, we are seeing an example of innovation realized to satisfy environmental/green needs. We can thus say that environmental/green needs have started to become so powerful that they drive innovation. However, reality is that environmental innovation has not yet reached the stage where it is able to change the lifestyles of citizens in the direction of low environmental impact.

9.4   Is Innovation Contributing to the Reduction of Environmental Impact?

Not all innovation taking place today necessarily contributes to the reduction of environmental impact. Both innovation that increases impact and one that lowers impact exist. Also, when we talk of a reduction in environmental impact, there are different degrees and ranges to which innovation contributes to such a reduction. In the last decade, global environmental issues have repeatedly been the topic of international conferences, corporations have thought up and implemented various solutions, many academic societies have been established, and scientific evidence of the deterioration of the global environment have been put forth. The environment surrounding innovation has, in this way, changed significantly, and citizens have come to express environmental/green needs driven by a "desire to contribute to the solution of environmental problems". The road to a true solution to the problem, however, is long and steep. Japan is said to have strong competitiveness in the environmental field, but how much has environmental innovation actually contributed to the reduction of environmental impact in Japan or in the world? Despite the fact that environmentally considerate products such as energy conserving appliances, solar energy, fuel cells, hybrid cars, electric vehicles, EcoCute, ecobags and many others have been launched, and despite the fact that Japan as a country has set CO_2 reduction targets, the actual emissions of CO_2 have increased 4.1 % between 1990 and 2010, and when looking at the household sector alone, the increase has been 35.5 % The industrial sector reduced its total emissions of CO_2 by 12.7 % in the same period, but this is the sector which with 35.3 % of Japan's total accounts for the largest portion of emissions, and thus Japan as a whole has not been able to reduce CO_2. Why is it, as this data shows, that Japan, despite various efforts to lower emissions, has not been able to achieve significant reductions? In Chap. 2, on eco-dilemmas, we looked at the essence of this situation, but, here, from an innovation angle let us contemplate what kinds of problems exist and where.

First of all, a factor that could be involved is that the diffusion ratio of environmentally considerate products is still low. If the ratio of diffusion in society remains low, the impact of innovation will not manifest itself in the form of CO_2 reductions for Japan as a whole, no matter how many energy and resource conserving products are developed.

Secondly, one could point to the fact that even if the energy efficiency of appliances is improved, the actual use pattern and time will lead to significant fluctuations in CO_2 emissions. Let us, for example, assume that the fuel efficiency of a hybrid car is twice that of a conventional gasoline car. If, however, the owner likes the comfort of driving in a hybrid car so much that he or she drives three times as much as previously, what will happen? Obviously, CO_2 emissions will increase due to the shift to a hybrid car. That is, depending on the use pattern of an energy conserving product, there is a possibility that it may generate a higher environmental impact than if the user kept the old device and was frugal with its use.

Thirdly, the fact that the reduction of environmental impact in individual devices is limited plays a role. If, as a result of technological development, CO_2 in a particular device is reduced in the order of a few percentage points, this cannot possibly have the cumulative effect of a reduction of several tens of percentage in society as a whole.

Fourthly, there is no guarantee that innovation would, of itself, help control or limit the excessive expansion of human activity. Innovation today is merely aiming for partial solutions, and thus, as both the needs of consumers and the development activity of engineers expand, we find ourselves in a situation where innovation itself is expanding. Unless we promote innovation under certain constraints aiming for total system solutions, we will be unable to control the expansion of human activity.

Finally, as a fifth factor, we can point to the fact that innovation to transform lifestyles in the direction of low environmental impact is not taking place. As long as we continue to use conventional types of products and services and maintain conventional lifestyles, we will be dependent on the extent to which impact is reduced in individual products or services, thus becoming unable to realize drastic reductions. This is also obvious when we consider that there are almost no products or services which are marketed with slogans such as "this product will change your lifestyle into one that is both wholesome and fulfilling and low in environmental impact".

At present, when it comes to the first factor above (low diffusion ratio) many initiatives are being introduced by national and local governments aiming to promote the diffusion of environmentally considerate products, and, gradually, initiatives and technologies to solve the second issue (use patterns) are being launched. When it comes to the third factor (insufficient impact reduction in individual devices), engineers have started noticing the existence of limits to reduction potential, but reality is that not much thought has been given to factors four (the role and direction of innovation) and five (the lack of lifestyle innovation).

9.5   Examples of Transitional Technologies

Aiming to deal effectively with the above factors four and five, the authors of this book developed the lifestyle design methodology described in previous chapters. This is a methodology which designs wholesome, fulfilling lifestyles while taking future, severe environmental constraints into account (Ishida et al. 2010). It is also a methodology for the planning of corporate strategy or policy proposals since it looks at the products, services, systems and policies required to enable the envisioned future lifestyle. Let us take a look at one example created with this methodology. In joint research between corporations and the Ecolab of the Graduate School of Environmental Studies Tohoku University, tests and demonstrations are being conducted on the theme of "Technologies utilizing weak energies". At the Graduate School of Environmental Studies, a project called "The development of energy conserving technology for ecohouses with storage batteries utilizing weak energy" was initiated, and technological development began in earnest in 2008, under the auspices of the Environment Ministry's scheme to develop technologies countering global warming.

The aim of this project is to create a technological platform which will help free people from a lifestyle in which they, unawares, consume huge volumes of energy. While maximizing the use of renewable energy, the aim is to collect so-called "weak energy" that is readily available but hitherto unused and store it in a storage battery in the house, thus enabling a shift to a lifestyle in which such energy is used in direct current form to power electric appliances (personal computers, TV-sets, LED, etc.). The ultimate goal is to enable a majority of Japanese households to shift to a lifestyle based on renewable energy.

"Weak energies" are the weak sources of wind power, hydropower, human power, gravity, pressure, solar power, etc., which are available in the vicinity of where people live. These sources of energy are not strong enough to be used directly for large electric appliances such as personal computers or TV-sets, but accumulating weak energy over time using storage batteries, the power becomes sufficient to be used in direct current form to operate such household appliances. Some examples include the energy of water flowing through drain spouts, the energy of water flowing from bath tubs to sewers, pedal power from an aerobics session in a sports center or at home or from a bicycle ride to and from school or work, the energy generated in revolving doors, the vibration energy generated when people go up and down stairs, the energy of a breeze going through the house, or non-conventional small solar panels that generate energy at lower voltage.

The concept of "accumulating weak energy" plays an important role exactly because the energy source is weak. First of all, in everyday life as large volumes of energy are consumed, Japanese often feel energy is too precious to waste, and the action of harnessing and storing weak sources of energy normally wasted may, thus, give people a sense of fulfillment. Also, people tend to be careful with the use of energy that has been gathered and stored with great care. Most people experience a sense of discomfort if the energy they themselves stored is used in a wasteful manner, and it is therefore likely that use patterns will also change. In this way, storing and

Generating, storing and using energy at home. From energy "lifeline" to "energy lifepoint"

Fig. 9.4 Enjoying life through the storing and use of weak energy

using energy that was so far wasted, leads to news ways of generating and utilizing energy. This is an effect one can expect to see exactly because we are talking of weak energy. Secondly, the very act of accumulating—or storing—energy in a storage battery is of importance. By accumulating the energy generated in a storage battery, energy takes on a more visible shape. It actually becomes possible to confirm with your own eyes how much energy it is that is wasted (Fig. 9.4).

In the above-mentioned project, we work on the technological and systemic development needed to bring the concept to life, conduct tests/demonstrations, and make proposals for lifestyles changes to the wider society. Most people realize that even if each individual source of energy is weak, by accumulating these they gain value as energy sources. Using your aero bike to generate such weak energy yourself; gaining energy from water flowing in your drain spout on rainy days; generating electricity from the breeze that goes through your house on a windy day or from your daily movements in and around the house—these and other approaches to technological development are being implemented by the project. By being aware of the energy source or the volume of energy from the very stage at which it is generated, many people agree that their way of thinking about energy, and thus lifestyles, changes.

In a survey we conducted on the issue of weak energy (Questionnaire on Ecohouses, n = 1,000) (Fig. 9.5), some unexpected findings were made. One such finding was that, when respondents were asked to identify "an unused energy source in their daily lives which they thought could have been harnessed", 42.1 % chose "energy from the pedaling of aero bikes/fitness bikes at home or at the training gym". This was the response getting the highest score; next came "energy from the flow of water through drain spouts" at 34.4 % and "the energy from the flow of water from bathtub to sewer" at 33.7 %

Also, when asked whether respondents would be interested in seeing and trying to utilize such unused energy sources, 77 % responded yes. 55 % responded that although each individual weak energy source is small, such energy would have use value when accumulated for utilization.

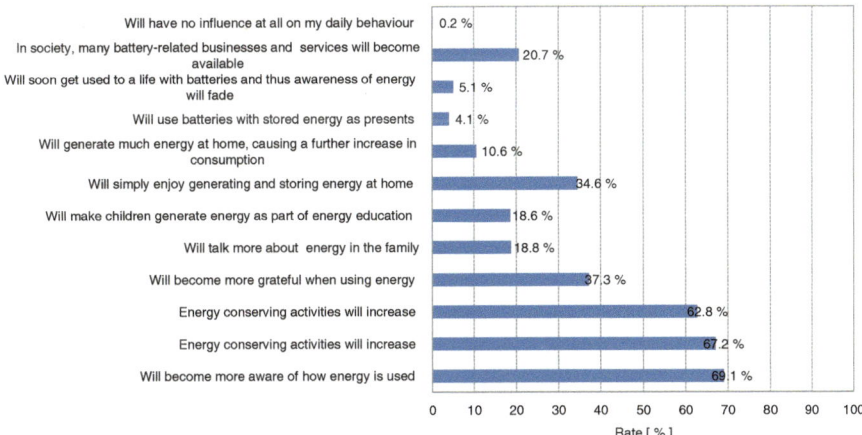

Fig. 9.5 Potential changes in consumer behavior (Tohji et al. 2011)

The system of generating energy from weak sources falls into two categories: one in which conscious action is taken to generate and store energy (for example, aero bikes/fitness bikes), and another in which no intentional action occurs (for example, flow energy from a drain spout). 53 % of respondents said they would be interested in using devices which store energy without the need for conscious action, while as many as 43 % said they would be interested in using both types of devices. There is a clear interest in society in generating and storing energy intentionally. Finally, respondents were asked whether they thought a shift from their present lifestyle, in which energy is not visible, to one in which energy is stored in a battery and then used in the house in direct current form—thus making energy more visible—would change their day to day actions. 30 % responded "I think it will cause drastic changes", while 57 % said "I think it will cause some small changes". That is, as many 87 % of respondents said this kind of visualization of energy in daily life would change their behaviour.

In more concrete terms, the changes imagined include "I will become more aware of the (the way I use) energy" (69.1 %), "Energy conserving efforts will increase" (62.8 %). Conversely, only a mere 0.2 % responded "It will not make any difference at all in my daily actions". Responses thus show that this kind of ecohouse system would change the ways in which people use energy. It points to the fact that such a system has the potential to transform the behavior of consumers.

The operational tests conducted by our Ecolab are still ongoing, but the evaluation of innovation in the context of transitional technology is actively being pursued today.

One of the unfolding experiments involves Shared batteries—the lifestyle with a social acceptability of 70 % described in Chap. 4.

In the old days in Japan, people would sometimes go to their neighbor to borrow *miso* or soy sauce, but this was not necessarily because they had run out of these two cooking ingredients. *Miso* and soy sauce here served as a communication tool, and people would exchange valuable information with their neighbors, such as

Fig. 9.6 Shared batteries

"You didn't look well yesterday, but today you seem to have recovered", or "You seem to have been worried about your daughter, but apparently the problem has been solved".

Today, what has taken on the role that *miso* and soy sauce used to play? Just as people gathered around the open air fire in the past, they will now gather around technology—the question is, what kind of technology? Using the backcasting approach, we arrived at the conclusion that one such technology is energy (Fig. 9.6). For example, people in a community might say, "Today we are having many guests coming to our house—could you lend us some energy, please?" or "We are going to be traveling for a while, you're welcome to use our energy in the meantime". The next step is to contemplate what technologies are required to enable this lifestyle of lending and borrowing energy in the community. Since we are talking of the occasional lending or borrowing of energy, a small system which includes a solar panel with a capacity of some 800 w—about 1/4 to 1/5 of the size of panels installed on house roofs today—and a small 1 kwH storage battery should suffice. In this system, energy is stored little by little and then used, and when insufficient, energy is bought from the power supply company (Fig. 9.7).

In operational tests made with this system, the energy consumption of the household, which was 10 kwh/day before the test, was reduced to 5 kwh/day. With a forecasting mentality, a 10 kwh/day consumption might, with the aim of increasing for example security, have increased by a factor of 1.5 to 15 kwh/day.

This is one example of how, even using technologies existing today, we can arrive at a fundamentally different solution by working with lifestyle design based on backcasting.

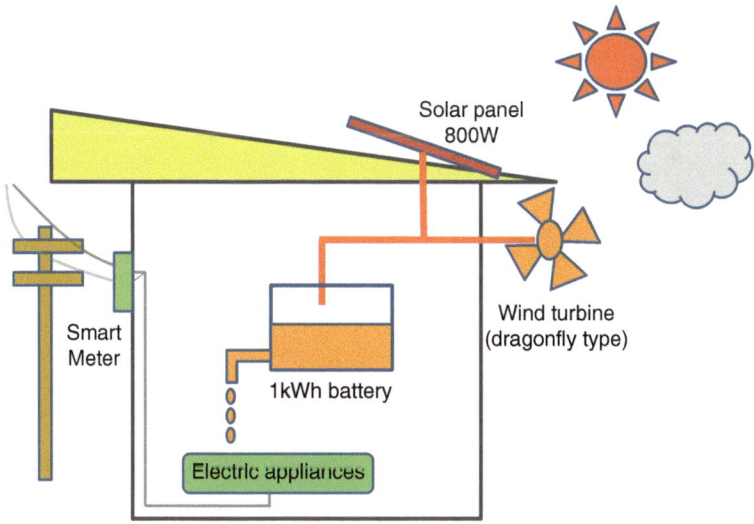

Fig. 9.7   Shared battery system

9.6   Smart Cities: Making New Lifestyle Proposals

In Akita City in northwestern Japan, a smart city project, based on the vision of "personalized ecology", has been launched. Presently, this is the first smart city initiative in the world which makes low impact lifestyle proposals. In this smart city project, participants can design future lifestyles with the backcasting methodology, and then go on to actually experience the life- or workstyles envisioned in their city. This city is not a showcase for low impact technologies, but rather a space in which new lifestyles can be experienced. Renewable energy is introduced to the largest possible extent, energy is shared, and based on the concepts of linking, enjoying and entertaining others (hospitality), the goal is to realize a city in which inhabitants can enjoy "personalized ecology" thus creating a new lifestyle brand.

As part of the Akita Smart City Project, we conducted interviews with nonagenarians. We interviewed elderly people in Akita City about pre-war ways of living in order to learn clever ideas and techniques for conserving energy and resources from that era. The next step, then, is to apply these ideas and techniques in future city planning in Akita City. We created a visual image entitled "a city continuing to co-exist with nature" (Fig. 9.8), which based on the application of pre-war ideas of living, shows a vision of a possible future way of living in the city. This image of the future helped encourage a lively and substantive discussion about the smart city project in the Akita Smart City Consultation Group. The point is that such a discussion needs to be not only about introducing new technologies, but must focus on the shape of future lifestyles in the city. Furthermore, the way this exercise made it possible to extract important elements relating to "Akita-likeness" was also noticed by participants. By using a methodology such as interviews with nonagenarians, who have

Fig. 9.8 A city continuing to co-exist with nature (futuristic vision of life in Akita City created from interviews with nonagenarians)

spent a lifetime in the city, it becomes possible to avoid the concern that smart city approaches might look the same for any city.

This project was initiated as recently as 2011 and will take some time to complete, but we are hopeful that these kinds of lifestyle-oriented smart city projects that help innovate lifestyles while incorporating local characteristics will spread to other places, both domestically and internationally.

9.7   Lessons of Community Design Learned from Interviews with Nonagenarians

Based on the interviews of nonagenarians in Miyagi Prefecture described in detail in Chap. 7, the authors of this book, in collaboration with the Sendai branch (Miyagi Prefecture) of Sekisui House, are contemplating how to establish a social platform supporting the transition to sustainable communities. This is one example in which we, based on hints gained in our interviews, are reshaping elements of pre-war living to fit a modern lifestyle. The vision is to establish fashionable spots in public spaced called "Parklets" for sharing renewable energy, thus also creating a space for people to gather and recreate while reviving the sense of community gradually being lost in modern society.

From our interviews with nonagenarians, we learned that in pre-war communities, people would share and co-manage resources needed for living, such as fuel, water, mountains, renewable energy and resources. People back then were not sharing everything, but they did share the truly important things in the community. And, since people realized that energy and resources were limited, the community had set up rules for the utilization thereof to avoid overuse. These were rules to maintain an abundant natural environment. This was an important element of life common to regions across Japan; a perspective on living which regarded constraints as essential to the forming of community.

We also know that in pre-war communities there were wooden "cool platforms" (rather like broad benches) where people would naturally gather, sense nature and relax while enjoying conversation with friends. People gained a sense of satisfaction sitting on these platforms even if nobody else was there, but when others came it became even more enjoyable and turned into a place for exchanging information. In the community there were always such public places for recreation.

Also, in the community, a proper system to educate children was in place. Children were not only brought up by their parents and grandparents; being praised or scolded by neighbors or people in the community were also important factors in their development. The bullies in the village would show younger children how to make tools to play and would even teach them about the ethics needed to live in a community. In everyday life, there were mechanisms to naturally pass on wisdom and techniques of living to the next generation. Moreover, through this process, people in the community were socialized, and we can safely say that it helped them gain a thorough understanding of each other's personality or life situation.

In the past, the lending and borrowing of everyday items frequently took place in the community, but according to the nonagenarians we interviewed, it was embarrassing, even in pre-war days, to borrow things. People did not share everything or lend and borrow carelessly. The lending party would be aware of the personality or life situation of the borrowing party, and thus with sympathy for the borrower's feeling of embarrassment, such lending and borrowing would be undertaken in a cheerful manner. This kind of relationship played an important role in the formation of community, particularly in Japan's unique culture where people want to avoid shame.

The "Parklet" spots for sharing renewable energy are designed around these concepts which form the backbone of community and function as venues through which new lifestyles outlined with the lifestyle design methodology can be enacted (Fig. 9.9). Parklets are set up in parks or other places in a town and provide a fashionable space for communication and recreation. It is a simple arbour with a table and chairs. The arbours all have different designs, and thus each has its own unique style. Local materials are used for at least part of the construction. Through the Parklet, people in the community can share energy. Parklets are equipped with recharging devices running on renewable energy (solar panels + a storage battery), and while relaxing, or letting the kids play in the park, people can recharge their mobile devices or personal computers. In the future, a new service, the donation of energy, will also become a standard function at the Parklet, and if energy is in

Fig. 9.9 Image of a Parklet

surplus it will be possible to donate this, and not money, to a local hospital or other places in need of energy. With this function, it is possible to make contributions to society through the Parklets set up all over Japan.

This is an initiative which, through the sharing of renewable energy, enables the networking in and of a community. Information platforms will also be added to Parklets making it possible to exchange, or work with, information. If the electricity generated and accumulated in the storage battery runs out, people can no longer recharge at the Parklet. This, however, helps teach people about the preciousness of energy. As we learned from the interviews about pre-war living, the establishment of appropriate rules are important for the Parklets to function properly. The cleaning of Parklets are undertaken by local residents, who are thus given a role to play in the community. If people start gathering at the Parklets, new business initiatives will probably also emerge. A Parklet can also become a medium for advertising. To be able to tell your friends about how to enjoy using a Parklet, it is possible to download applications at each different Parklet explaining the particular functions available at that location. As the number of Parklets increases, business chances will be created thanks to a network effect. We could call such Parklets a new, real-world social network helping people to gain a sense of wholesomeness even living under severe constraints. Through such design, it is possible to present new lifestyles which pass on the wisdom of pre-war living.

The use of Parklets is not restricted to local residents. People will even travel to find these fashionable spots, and some take photos in front of them. Friends gather and a vibrant exchange takes place also with people from other regions. People can use recreation time at these Parklets to donate energy which is then used, for example, as an emergency power supply of a hospital of other public facilities; in this way it is possible contribute to society and, instantly, gain a sense of fulfillment. As they become able to use renewable energy in everyday life using the Parklets, more and more people get knowledgeable about the handling of such energy sources—and through this, a shift to a lifestyle with a greater awareness of the importance of energy becomes more likely. When natural disasters occur, the battery installed at the Parklet is used as an emergency power supply by the local residents themselves.

Can we, perhaps, hope for these small and fashionable Parklets to help revive a traditional Japanese sense of community in co-existence with nature? The challenge of trying to spread such Parklets across, primarily, the Tohoku Region—Miyagi, Akita, Iwate and other prefectures—has just begun.

Bibliography

Akita City (2011) Akita-shi sumuaato shiti purojekuto kihon keikaku (Akita City Smart City Project – Master Plan), Japan

Furukawa R (2010) Kankyou Seiyakuka ni okeru inobeeshonryoku wo mochihajimeta kankyou niizu (The innovation power of environmental needs under environmental constraints), Tohoku University Press, Miyagi

Furukawa R, Hiroaki S, Ishida H (2008) Change of eco-innovation in energy consuming products industries in Japan. In: 21st international CODATA conference scientific information for society – from today to the future, Kyiv, p 111

Ishida H, Ryuzo F, Dentsu Grand Design Laboratory (2010) Kimi ga otona ni naru koroni. Kankyou mo hito mo yutaka ni suru kurashi no katachi (By the time you grow up – a way of living that makes both the environment and people rich). Nikkan Kougyou Shimbun, Tokyo

Ito K, Furukawa R, Sasa H, Ishida H (2009) Reizouko no kankyou inobeeshon no nichibei hikaku (Japan-US comparison of environmental innovation of refrigerators). The Japan Society for Policy and Research Management. 24th Annual Academic Meeting, Lecture summary, Japan, pp 657–660

Tohji K (2011) Sumaato hausu no hatsuden, chikuden, kyuuden gijutsu no saizensen (The frontiers of energy generation, storing and supply technologies for smart houses). In: Furukawa R (ed) Chikuden kinou tsuki juutaku no kaihatsu (The development of houses with built-in storage batteries), vol 5, Chap 1. CMC Books, Tokyo, pp 253–265

International Energy Association (2008) World Energy Outlook 2008. International Energy Association, Paris

Chapter 10
Manufacturing That Takes Nature into Account: The Shape of Nature Technology

Abstract Envisioning wholesome, fulfilling lifestyles full of excitement even in the face of the severe environmental constraints of 2030, then searching in nature for the technologies required to make with vision come true—this is the new approach to manufacturing: Nature Technology. Why search for answers in nature? The reason is twofold: one is that nature, through a continuous process of natural selection, has achieved the optimum form of evolution adapted to the various local and global environments through time and has done this in a system of perfect cycles driven with a minimal input of energy. This is indeed the "sustainable society" which humanity, despite lofty ideals, is still far from achieving—created by nature with the use of solar energy and the abundance of things found on Earth's surface. We, humans, need to re-learn the mechanisms and systems—as well as the social process of creative destruction—found in nature. A second reason is that the modern technology created as a result of the industrial revolution starting in Great Britain succeeded only by removing itself and thus humans from nature, and as a result thereof is intricately linked to the emergence of global environmental problems.

Keywords Biomimetics • Materialistic civilization • Mimicking function • Mimicking shapes and forms • Mimicking systems • Nature mimicking • Nature Technology • The idea of man conquering nature • The industrial revolution

10.1 Nature Technology: A New System of Manufacturing

To be able to pass on the blessings of Earth's finite resources to the next generation, we urgently and thoroughly need to reconsider the relationship between the global environment and human beings. The time has come to create a new mould for manufacturing and living which acknowledges the reality that we, humans, possess "genes of desire"—that is, we do not willingly let go of modern conveniences and comforts once gained—and which, with the smallest possible impact on the global environment, responds to this reality.

E.H. Ishida and R. Furukawa, *Nature Technology: Creating a Fresh Approach to Technology and Lifestyle*, DOI 10.1007/978-4-431-54613-9_10, © Springer Japan 2013

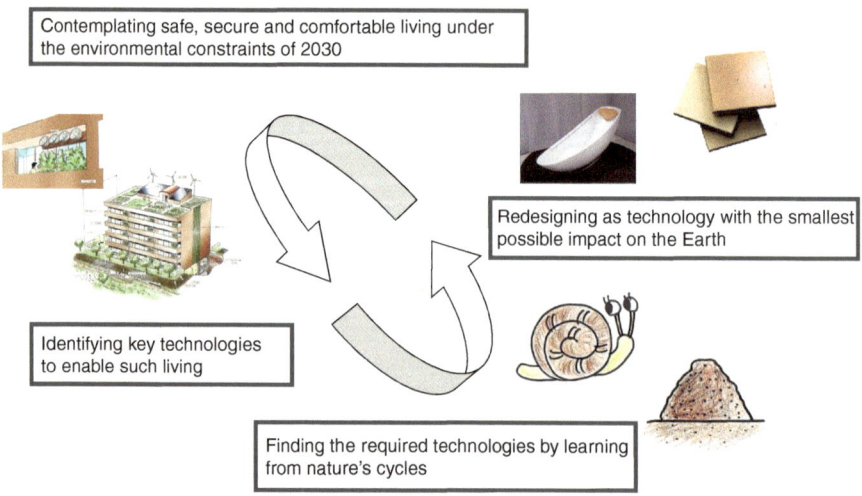

Fig. 10.1 A system for the creation of nature technology

Nature Technology is one such response; a new system of manufacturing which intelligently utilizes the amazing powers of nature. Nature Technology does not, as is the case in a conventional approach to technology, create things that are not found on the surface of the Earth; it is, rather, an entirely new way of approaching manufacturing and living which looks at nature—functioning in perfect cycles and with a minimal input of energy based on 4.6 billion years of natural selection—from a scientific viewpoint and redesigns it as new forms of technology.

A virtually infinite number of living beings exist in nature. Nature does not use oil or other fossil fuels to support these different forms of life. Mainly, all living things on Earth are connected through the clever utilization of abundant energy reaching the Earth from the Sun. Or, the great forces of the Earth, such as tectonic shifts and diagenesis turn organic life forms into resources and energy. In nature, we find an astonishing (to us humans, that is) number of fascinating seeds of technology. Nature's cycles are the fruit of wisdom developed through life's 4.6 billion year history; Nature Technology is a new attempt at using the skills we human beings have acquired as tool makers and users to learn from, and bring to life, these amazing powers of nature.

Today we must, through the intelligent use of technology, make new proposals for ways and forms of living. Technology is needed that can make statements about future forms of living. These are also, we may say, the kinds of technology that—as is required in our age—assume responsibility for lifestyles.

As its point of departure, Nature Technology (Fig. 10.1) learns from nature's perfect cycles driven with a minimal input of energy. However, this does not mean merely substituting a technology learning from nature for a conventional technology. It must, also, involve a shift to a way of living in which we do no longer think of, for example, an air conditioner as a "thing/device", but rather focus on the "action/function" of controlling indoor air quality. In this way, we may satisfy people's desires for convenience and comfort—once gained so difficult to let go of—not in a materialistic way, but rather by fulfilling immaterial, or spiritual, urges. In more concrete terms, we must, as previously described, envision wholesome, fulfilling lifestyles under the severe environmental constraints of 2030 using the backcasting methodology and then distill the technologies required to make this vision come true. The important point is to take lifestyles as the point of departure for this process. This is fundamentally different from a conventional approach to technology development. In conventional development methodologies, a forecasting approach is most often taken, which first creates an image of the technology and then considers how to introduce it into a lifestyle. This approach cannot break free from wasteful thinking based on the notion that "to live humane lives, people must be materially prosperous", and thus ends up triggering eco-dilemmas. It is crucial that we first use backcasting to create a clear picture of a lifestyle, and then contemplate the technologies needed. This is the only way to develop technology which assumes responsibility for lifestyles while also ensuring that the prosperity needed to live fulfilling lives is secured.

When the elements of technology to aim for have been clarified, the next step is to search for these in nature. Obviously, this is, as mentioned above, important because nature has created perfect cycles with minimal energy inputs, but furthermore, there is a need to incorporate a view of nature into technology itself.

10.2   Can We Learn from Nature?

We can learn two major lessons from nature. One is the fact that nature, through a continuous process of natural selection, has achieved the optimum form of evolution adapted to the various local and global environments through time and has done this in a system of perfect cycles driven with a minimal input of energy. This is indeed the "sustainable society" which humanity, despite lofty ideals, is still far from achieving—created by nature with the use of solar energy and the abundance of things found on Earth's surface. We, humans, need to re-learn the mechanisms and systems—as well as the social process of creative destruction—found in nature. A second important lesson we can learn from nature is that modern technology, created as a result of the industrial revolution starting in Great Britain, succeeded only by removing itself and thus humans from nature, and as a result thereof is intricately linked to the emergence of global environmental problems.

10.2.1   Nature Drives Perfect Cycles with a Minimal Input of Energy

In Nature Technology, the idea of mimicking nature is of great importance. The notion of Biomimetics—the science of mimicking living things—was first proposed in the 1950s by neurophysiologist and inventor Dr. Otto Schmitt. Based on research of the nervous system of squids, Schmitt in 1934 invented the so-called Schmitt

Fig. 10.2 A shark's dermal denticles (*left*, Photo: Sue Lindsay © Australian Museum), the kingfisher (*right*)—a bird able to dive into water at top speed

Fig. 10.3 The water repellent lotus leaf is covered by small protuberances (Photos: © Dr. Osamu Takai, Nagoya University)

Trigger, an input circuit system for digital circuits which arguably is the foundation for present electronic devices. Of course, centuries before this, Leonardo da Vinci left us many sketches which, based on the mimicking of a bat's flying skills, show how to fly through the air, and finally in 1903, the Wright brothers famously completed their first flight (For quite a while after that, people would argue that "machines cannot possible fly through the air" and refused to acknowledge their great achievement).

There are three large categories when we talk of Biomimetics. The first is the mimicking of shapes and forms. The skin of a shark is covered by tiny protuberances with a tooth-like structure called dermal denticles, and these help lower low water resistance making it possible for the shark to swim fast.

By mimicking this mechanism, swim wear with reduced water resistance was developed. Mimicking the water repelling mechanism of a lotus leaf helped give birth to external wall material that doesn't stain easily, and the shape of the kingfisher's beak was inspiration for the shape of the front car of the 500-series Shinkansen Bullet Train (Figs. 10.2, 10.3).

The second category is the mimicking of function. Insects do not have an immune system such as that of humans, but when facing an attack by bacteria, they immediately defend themselves by making a protein in their bodies called defense-related protein. From this protein, an anti-cancer drug and other medicinal products are

Fig. 10.4 A school of bigeye jackfish (Photo: © Sea Dream Okinoerabu)

about to see the light of day. Research is also underway on a material that, using the mechanism of leaves dropping from deciduous trees, disintegrates and falls apart once its lifetime has been reached.

Finally, there is the mimicking of systems. How do schools of fish form, and why do the fish not collide with each other in such schools? Why are birds able to fly in beautiful formations? Understanding such systems may help us change traffic systems or information networks (Fig. 10.4).

An even broader concept than Biomimetics is nature mimicking. For example, mimicking the mechanism behind the way in which hot magma cools and turns into volcanic rock is the idea behind ceramics. Nature—including both below and above the Earth's surface—has created large and perfect cycles. Looking at these through the lense of global environmental issues and ways of living is the essence of Nature Technology.

Let us take a look at some concrete examples of this kind of technology.

If we look, first, at Nature Technology that contributes directly to solving global environmental issues, the honeycomb structure of a bee nest is a good example—it is light but of great strength and is used as structural material in airplanes, boats and residential houses (Fig. 10.5). Pine cones will close when moist content is high and open up when it is low in order to allow for the seeds kept inside to disperse. Based on this principle, smart clothing which automatically regulates temperature was developed. When moisture content in the air is high, the gaps between fibers will open up, and when moist is low, they will contract, thus regulating temperature.

Fig. 10.5 The honeycomb structure of a honeybee nest

Fig. 10.6 The hump of the humpback whale

The humpback whale, as the name suggests, has humps on its fins (Fig. 10.6), and research has revealed that these humps reduce water resistance significantly. A prototype wind generator applying this structure of a humpback whale fin has been developed. When it comes to electricity generation, owls also make a contribution. The flight feathers of an owl have ragged, saw-like edges called serrations which

Fig. 10.7 Dragonflies have rugged wings

eliminate the sound of flight. This mechanism was applied in pantographs (current collectors) for the 500-series Shinkansen Bullet Train, resulting in a noise reduction of more than 30 %. Since sound is energy, less noise means that energy is being used more efficiently, and for this reason, the same mechanism is being studied for the potential use in wind generators.

In the sea, seaweed rolls gently with the movement of the sea, and a generator utilizing such gentle flow energy to generate electricity has also been manufactured as a trial. Furthermore, amazing findings have been made recently in the research of dragonflies. The wings of a dragonfly are not, as with eagles or pigeons, smooth, but have a rugged structure. Dragonflies use this to fly skillfully even when there is no wind (Fig. 10.7). Actually, the mechanism behind this is not yet fully understood, but researchers have discovered that the rugged shape of the wing creates air turbulence (much like whirlpools) which creates an effect similar to that of a ball bearing. It appears that this allows for air to be transported over these "bearings" as if on a conveyor belt, thus giving the dragonfly its needed uplift. Using this mechanism, it is possible to make a wind generator that generates energy even in a slight breeze (for details, see Chap. 11). Seeds of the tropical plant Alsomitra will fly long distances in tropical rainforests with no wind, and their shape has been adopted in hang-gliders and, in the past, in airplanes (Fig. 10.8).

Velcro tape, known all over the world today, is a registered trademark of Japanese chemicals company Kuraray. The Swiss national George de Mestral gained hints from the cockleburs sticking to his clothes as he went hunting and first produced Velcro commercially in 1955. Cockleburs have numerous small hook-like structures on their surface (Fig. 10.9) which hitch on the fiber of clothes or a dog's fur. De Mestral reproduced this mechanism using a special type of nylon. Kuraray bought

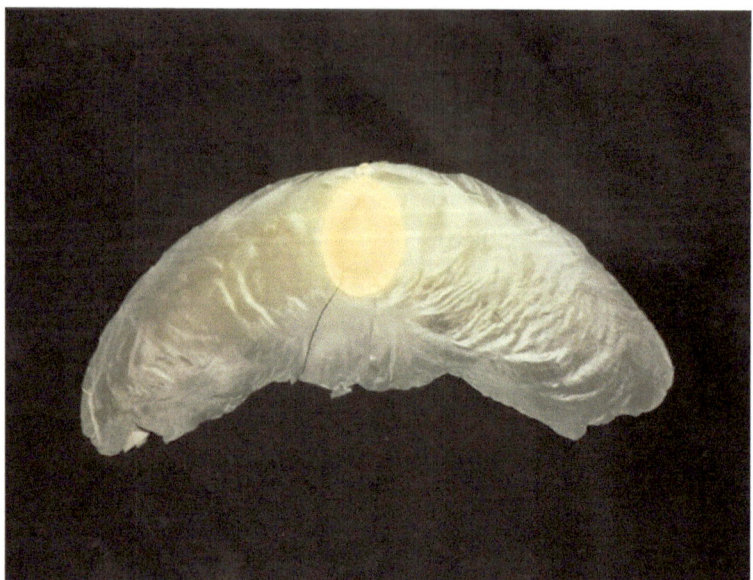

Fig. 10.8 A seed from the Alsomitra plant

Fig. 10.9 The cocklebur which led to the invention of Velcro

the patent of the invention in 1964, and after being used to cover the headrests on the Shinkansen Bullet Train starting operations in the same year, the utilization of Velcro expanded rapidly.

Due to the effect of population increase and global warming, a scramble for water has already begun. Here, the unique ways of collecting water mastered by the fog

Fig. 10.10 The fog
collecting ground beetle
which collects water from the
air with its body

collecting ground beetle living in deserts may come to be of help. In the Namib
Desert in Africa, fog covers the landscape every morning. The fog collecting ground
beetle does not waste this opportunity; facing the wind at just the right angle, it turns
upside down with its back facing in the direction of the wind. On its back it has
numerous hydrophilic protuberances with which it attracts the moist in the fog.
This moist gathers and gradually turns into droplets of water. As the weight of the
droplet increases, they run down hydrophobic channels between the protuberances
and enters the beetle's mouth (Fig. 10.10). Experiments are underway which,
using this mechanism, set up hydrophilic and hydrophobic areas on the surface of
buildings with the aim of making water from air.

Cellulose—a type of carbohydrate found in wood and fibre—is a tricky substance,
indigestible not only to humans, but to all animals except termites. It happens,
though, to be about as strong as iron. It is actually cellulose which creates the unique
jelly-like texture of *nata de coco* (known from Filipino desserts). In *nata de coco*,
the cellulose is distributed in a mesh-like pattern, and special filters have been
created by removing the 99 % water of which *nata de coco* is composed and utilizing the 1 % that is cellulose. Since the diameter of *nata de coco* cellulose is smaller
than the wavelength of visible rays of light, it looks transparent to the human eye
(Fig.10.11). Flexible organic electroluminescence displays that can be made by
inserting synthetic resin into this mesh-like cellulose are also catching the
world's attention.

Another example comes from the bottom of the sea. In the deep sea, numerous
spots called black smokers have been discovered from which water heated by
magma gushes out together with hydrogen sulfide and heavy metals. Near these
spots, shells called calyptogena have been found in large numbers. With the help of

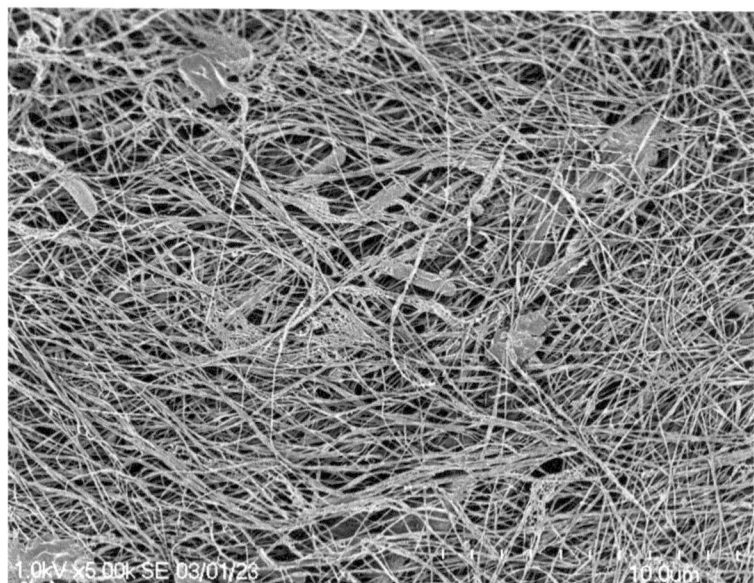

Fig. 10.11 A microscopic image of the tiny cellulose fibers found in *nata de coco* (Photo: ©
Dr. Hiroyuki YANO, Research Institute for Sustainable Humanosphere, Kyoto University)

sulphur bacteria, the shells transform hydrogen sulfide—a deadly poison to
humans—into energy. It is conceivable that if we could learn to utilize this mechanism,
a new source of energy could be created.

Yet other examples include new approaches to adhesion learned from geckos or
from a bacteria called caulobacter crescentus; underwater adhesion learned from
blue mussels or barnacles (Fig.10.12), a relative of crabs and lobsters; spider thread
as soft as silk yet as strong as steel; the shell of an abalone so strong that even a
hammer cannot break it; the ability of the morpho butterfly to generate colours
without pigments; the ability of the sandfish to move through desert sand as if it
were swimming; the skills of a cupedidae beetle to detect a forest fire from several tens
of kilometers away; the ultrasensitive sensing ability of a cricket which allows it to
sense predators from movements of the air, and so on. Numerous attempts at developing technology that harnesses the ingenious mechanisms found in nature are taking
place today. For more details and examples, see, for example, our website *Sugoi
shizen no shouruumu* (A showroom of nature's wonders) http://nature-sr.com/.

Here we have touched on but a few examples, but nature, in this way, does not
need to use scarce resources such as rare metals or rare earth; it produces multitudes
of life forms at normal temperature and pressure with materials found in abundance
aboveground and creates perfect cycles. It is no exaggeration to say that nature is a
treasure house of technologies. However, we have unfortunately only rarely, so far,
studied the mechanisms and systems maintaining nature's cycles from the viewpoint
of engineering. In conventional engineering, the main approach has been to work

Fig. 10.12 Barnacles—capable of underwater adhesion

with underground resources or fossil fuels at high temperature and high pressure to make things not found in nature, and only very rarely have attempts been made to look at nature's cycles or natural functions from an engineering perspective. Therefore, in order to promote such an approach, the creation of a database with applicability for engineering is much needed, but the conversion technologies needed to make the huge existing volumes of data on biology applicable to the world of engineering are only now entering a study phase, and we can, thus, hardly say that actual ways of applying such technologies have been established.

Let us at take a look at trends in different countries around the world. In October 2010, the 10th meeting of the Conference of the Parties to the Convention on Biological Diversity (COP 10) was held in Nagoya, Japan, where the main topics discussed included post 2010 targets for the protection of biodiversity, an international agreement on the access to and fair and equitable sharing of the benefits arising from use of genetic resources, and operational rules for the principles of liability and redress adopted in the previous Cartagena Protocol. When it came to specific detail, however, the participating countries prioritized national interest—much in the same way as with climate change negotiations—and no major progress was achieved. Preceding this COP10 conference, the German government in 2009, as a follow up to COP9 held in Bonn in 2008, initiated a Business and Biodiversity Initiative dubbed "Biodiversity in Good Company". The website of this initiative includes examples of biomimetic technologies that have already been developed and marketed. Also, Germany through financial support helped establish the Biomimetics Association and is vigorously promoting cross-disciplinary academic research in the field. Apparently, as part of national strategy, Biomimetics is being supported as a new field with future potential. In October 2012, with Germany as a driving force, the first meeting of a process to establish ISO standards for Biomimetics was held (ISO/TC266 Biomimetics).

In the United States, science writer Janine Benyus in the latter half of the 1990s put forward the idea of "Biomimicry" which, with her publication of "Biomimicry—Innovation Inspired by Nature" rapidly caught on. Benyus and others later established a non-profit organization, The Biomimicry Institute, and a consulting service, The Biomimicry Guild, and from these efforts a new industry is beginning to take shape. Toward the end of 2010, an economic evaluation of such efforts was published which estimated that biomimicry has the potential to become a USD 300 billion industry creating 1.6 million new jobs within 15 years.

In England, the BIONIS Network, with corporations and universities as main participants, was established in 2002. From around 2000, the number of academic papers on Biomimetics rapidly started increasing and by 2012 had reached more than 1,600, thus indicating that a new trend is being created.

In Japan, the Ministry of Education, Culture, Sports, Science and Technology (MEXT) as part of its twenty-first century Centers of Excellence initiative has adopted programs concerning "Biomimetic Manufacturing" and "New Uses of Biological Resources", and in the "Declaration on Biodiversity", launched in 2009 by the Keidanren (Japanese Business Federation), article 5 mentions "We will aim to create new industries, ways of living, and culture inspired by biodiversity" and goes on to explain this in detail in the following way: "Learning from the profound and boundless blessings of nature, as well as from tradition and the wisdom of our forefathers, we will, through the advancement of corporate activity with low impact on biodiversity and the development of environmental technology, aim to create sustainable industry, ways of living, and culture". Meanwhile, the corporate world has also launched some related products. Comparing with other nations, however, in which Biomimetics is regarded as a whole new academic field and where scenarios are being pursued to use it as a trigger for the creation of new industries, Japan is still lagging behind, and it seems to us that actions taken have been rather haphazard and disjointed. Against this background, there are recent positive recent developments. In 2012, a program called "Innovative Materials Engineering Based on Biological Diversity" (Chair: Masatsugu Shimomura, Tohoku University) was started as one of the new academic fields under the auspices of MEXT, and in 2009, The Nature Technology Research Consortium (Coordinator: Emile H. Ishida, Tohoku University) began activities as a research association under the "Japan Manufacturing Conference" (MONODZUKURI. Nippon. Conference). The former works mainly with scholars, the latter with people from the corporate world, but the purpose of both initiatives is the same and through the cooperation of these two groups, it is our hope that Biomimetics research in Japan, somewhat belatedly, will now pick up speed. Taking environmental constraints into account and pursuing the issue of lifestyles are new values we hope to add to establish an innovative new system for technological development. In any case, we strongly hope that a country like Japan, which has manufacturing as one of its foundations, will urgently, and as a national strategy, pursue new kinds of manufacturing which intelligently harness the amazing powers of nature. It is in this context that the Ministry of the Environment decided in April 2013 to initiate a policy planning and institutional design program which—with the concept of "a civilization of environment and life" as its point of departure—works towards the creation of a sustainable society in which greater emphasis is placed on spiritual/immaterial fulfillment than on materialistic prosperity.

10.2.2 The Success of the Industrial Revolution Was Based on the Principle of Humanity's Separation from Nature

Another important thing we can learn from nature is the way in which the success of the industrial revolution was based on our separation from that very nature. Why did the anthropocentric technology which has invited a civilizational crisis develop to the extent we see today? And, what do we have to consider today in order to change the direction of technological development away from material desires towards the encouragement of spiritual or immaterial desires? To pursue these issues, we need to take a look at the development of modern technology in Europe which led to the birth of our present underground-resource based civilization. From the eighth until the thirteenth century, Europe was in what is called the medieval warm period. Agriculture spread across the Alps into Western Europe, where technological innovation to utilize water and wind mills extensively continued. These technologies gradually became essential to the textile and steel industries and, as we shall see later, were an important factor in setting the direction towards the modern, Western European civilization in which man aims to conquer nature. The cooling of the European climate starting in the latter half of the thirteenth century, with a small intermission in the sixteenth century, reached a peak in the second half of the seventeenth century. Harvests in Europe around that time dropped drastically and the cooling triggered major famines. Around the year 1500, when the small ice age was at its height, the climate not only reduced agricultural productivity, but also caused the spreading of pests which, combined with the effect of numerous wars, led to sharp decreases in population. European society at that time, it is said, was on the verge of collapsing. On the other hand, a lack of labor led to the growth of livestock and dairy farming, and machines were introduced to make up for the lack of hands in agriculture and industry. Gradually, the idea that man was part of nature faded, humans became increasingly removed from nature and, attempting to alter and reorganize nature artificially, aimed to become its conquerors. Such developments led to, and were further fueled by, Francis Bacon's concept of human dominance over nature and Rene Descartes' mechanistic worldview. This gave birth to a modern, Western philosophy justifying modern civilization and, in a concrete way, helped build a human empire enslaving nature. The philosophy of dominance over nature advocated by Francis Bacon expresses itself in his philosophical work "Novum Organum", where he—while acknowledging the overwhelming power of nature—at the beginning of volume one, writes, "Nature cannot be conquered but by obeying her". Rene Descartes, by contrast, was a skeptic who based his philosophy on the statement "Cogito ergo sum"—I think, therefore I am. Nature was seen as existing in opposition to man, and thus, the better human beings understood the laws of nature, the easier it would become for them to control and manipulate nature at their will.

Such a mechanistic worldview became the most fundamental principle of modern science and technology, and as a result thereof a convenient modern society was built enabling the endless expansion of material desire. These ideas became so dominant across the world that even non-Western countries could not develop without adopting Western civilization. If attempts were made to reject this civilization, countries would be colonized, and, thus, regardless of whether or not there was any willingness to adopt Western thought, the world was gradually swallowed up by Western civilization. This anthropocentric worldview, however, does not allow for any restraint in relation to nature, and thus led to the creation of severe environmental constraints.

A dualistic worldview developed in order to understand the world in this way ended up creating a dichotomy between humans and the natural world. Of course, Cartesian thought did not suddenly trigger modern civilization. Descartes perfected mind-body dualism, but it is said that such dualism is found even in the works of Plato or Aristotle—major foundations of Western thought—or in Hinduism.

Human dominance over nature was, thus, not just an idea arising from the works of Bacon and Descartes. In the Old Testament, Genesis 1:26 states, "We will let them rule the fish, the birds, and all other living creatures". The Christian view of nature, which took shape long before the scientific revolution, looked at human beings as being in opposition to nature. Francis Bacon's thoughts on human dominance over nature derived from an objectification of nature for scientific investigation, thus creating an intrinsic opposition between human beings and nature, and Descartes took this worldview even further. When did such views of nature originally start taking shape? According to Japanese philosopher Takeshi Umehara, they had their roots in the civilization existing in the Western part of the Eurasian continent where wheat and dairy farming formed the foundations of society. The first nation (city state) was created by the Sumerian king Gilgamesh (Gilgamesh is regarded as the historical king of the Sumerian city state Uruk, in modern day Iraq, around the year 2600 BCE). According to the Epic of Gilgamesh, the king went on a journey to search for cedar trees not found in Mesopotamia. The cedar forest was, legend tells, protected by the guardian spirit Humbaba, but turning his back on the gods, Gilgamesh killed Humbaba and brought home cedar trees to Uruk. In this way, after unifying the city state of Uruk, the first undertaking of Gilgamesh was to kill the guardian of the forest, Humbaba. According to Umehara, allowing the murder of the forest's guardian spirit is a train of thought found in the Sumerian civilization—which came to form the foundation for Western civilization—and similar thinking was later reflected in the works of, for example, Plato or Descartes, gradually becoming a basic principle of Western civilization.

Did the idea of human dominance over nature then begin in this era? Apparently not; our species, homo sapiens sapiens, originated in Africa some 200,000 years ago, and some 40,000–80,000 years ago had spread to the entire African continent, in the process eradicating some 40 % of large mammals such as great wild boars and wild stallions. Some 30,000 years ago, after our species had spread to Europe, 50 % of great mammals such as mammoths and bison had disappeared. 15,000 years ago, the warm period following the Ulm glacial stage (the last glacial period) caused the Alaskan glaciers to melt, and with this, humanity rapidly spread southwards, within 1,000 years reaching the Southern tip of South America. In this time frame, humanity caused the eradication of some 73 % of herbivorous animals in North America above 50 kg, and for South America, the figure was as high as 80 %. There are locations where it appears that humans, almost as part of a game, hunted and forced hundreds of mammoths over cliffs. As part of our DNA, we humans have genes of desire which enable us to conquer and dominate nature without moral restraint in order to satisfy our own greed. Apparently, a philosophy of human dominance over nature, and the existence of genes of desire, are some major characteristics of Homo sapiens sapiens.

Nature Technology, which harnesses the amazing powers of nature intelligently, is a process which first envisions lifestyles with a backcasting mentality, then distills the technologies required to realize these, search for hints and ideas for technologies in nature, and finally, through the filter of sustainability, designs new technologies. In lifestyle research we have learned that people strongly yearn for nature, and that numerous seeds of the new technologies we need are hidden in nature. The Spanish architect Antonio Gaudi once said, "Human beings cannot create anything, they can only discover". Nature is a textbook in which all the answers are written. Why, then, did technologies separating humans from nature come to rule the world and lead to environmental degradation? This, of course, is due to the nature of human arrogance which led us to believe we can control nature as our slave, and our predicament was caused by the illusion that underground energy and mineral resources could be a realistic solution to meet our desires. If we continue as presently, will future generations, when trying to unravel the "picture scroll" of Earth's 4.6 billion year history, be able to find in it an era, as thin as a thread of hair, with immense impact on the future of life on the planet? This thread is the twenty-first century, in which humanity will have used up all underground energy and mineral resources.

Let us reconsider where we stand and what we can do. What would a technology be like that does not separate itself from nature? What would an industrial revolution look like that does not separate itself from nature? And would such technology and industry be able to avoid causing environmental degradation?

Bibliography

A showroom of Nature's Wonders. http://nature-sr.com/
Akaike M (2006) Konchuryoku (Insect power). Shogakkan
Bacon F, Juichi K (transl.) (1978) Novumu oruganumu – shinkikan (Novum Organum). Iwanami Bunko, Tokyo
Benyus J (2002) Biomimicry: innovation inspired by nature, 2nd edn. William Morrow Paperbacks, New York

Descartes R, Tanigawa T (transl.) (2002) Houhou josetsu (Discours de la méthode pour bien conduire sa raison, et chercher la vérité dans les sciences). Iwanami Bunko, Tokyo

Forbes P (2006) The Gecko's foot: bioinspiration: engineering new materials from nature. W. W Norton & Company, Tokyo

Fijisaki K (2010) Konchuu miraigaku (Futurology of insects). Shinchosha, Tokyo

Heidegger M, Hara T, Watanabe J (transl.) (2003) Sonzai to jikan (Sein und Zeit). Chuo Koronsha, Tokyo

Honda H (2010) Katachi no seibutsugaku (A biology of forms). NHK Publishing, Tokyo

Ichikura H (1986) Gendai furansu shisou e no osasoi (An invitation to contemporary French thought). Iwanami Shoten, Tokyo

Ide T (Supervisor), Mitsuo Y (ed) (1968–1973) Arisutoteresu zenshuu (Complete works of Aristotle), vols 3, 6, 12, 13. Iwanami Shoten, Tokyo

Ishida H (2009) Channeling the forces of nature. Tohoku University Press, Miyagi

Ishida H (Supervisor) (2011) Sugoi shizen zukan, (Amazing Picture Encyclopedia of Nature). PHP Publishing, Tokyo

Ishida H (2010) Atarashii kurashi to tekunorojii wo kangaeru iinkai (Committee on New Ways of Living and Technology) Chikyuu ga oshieru kiseki no gijutsu (Miraculous technology the Earth teaches us). Shodensha, Tokyo

Ishida H, Shimomura M (2011) Shizen ni manabu neichaa tekunoroji (Nature technology learned from nature). Gakken Publishing, Tokyo

Ishida H, Matsuda M, Eguchi E, Nishizawa M (2011) Yamori no yubi kara fushigi na teepu (Wondrous tape from the foot of the gecko). Arisu-kan, Tokyo

Kamiyoshi S (2004) Beekon zuihitsushuu (Recollections of Bacon). Iwanami Bunko, Tokyo

Kitano H, Takeuchi K (2007) Shitataka na seimei (Life is wily). Diamond Publishing, Tokyo

Kobayashi M (1995) Dekaruto tetsugaku to sono shatei (The philosophy of Descartes and its reach). Iwanami Shoten, Tokyo

Koizumi K (2007) Kikou hendou to bunmei no seisui (Climate change and the rise and fall of civilizations). J Geogr (Chigaku Zasshi) 116:62–87

Kurokochi S (2000) Kindaisangyoushugi no kigen – furanshisu beekon zou no saihyouka (The origins of modern industrialism – a reevaluation of Francis Bacon). In: Socio Science, Japan, pp 263–275

Maruyama S, Isozaki Y (1998) Seimei to chikyuu no rekishi (A history of life and the Earth). Iwanami Shinsho, Tokyo

Nagashima T (2007) Ka ga noukousoku wo naosu! Konchuryoku no kyoui (Mosquitoes cure brain infarction! The marvels of insect power). Kodansha, Tokyo

Nagayama K (1995) Jikoshuuseki no shizen to kagaku (Self-aggregating nature and science). Maruzen, Tokyo

Nakajima Y (1995) Tetsugaku no kyoukasho (A textbook of philosophy). Kodansha Tokyo

Nakamura K, Matsubara K (1990) Seimei no sutoratejii (The strategies of life). Iwanami Shoten, Tokyo

Okada A, Kobayashi T, Takahito M (2000) Kodai Mesopotamia no kamigami – sekaisaiko no ou to kami no kyouen (The gods of ancient Mesopotamia – a symposium between the world's oldest king and god). Shueisha, Tokyo

Shimomura M (Supervisor) (2011) Jisedai baiomimetikkusu kenkyuu no saizensen (The frontier of next generation biomimetics research). CMC Books, Tokyo

Stewart I (1997) Nature's numbers: the unreal reality of mathematics. Basic Books, annotated version, Tokyo

Tanigawa T (1995) Dekaruto kenkyuu – risei to kyoukai to shuuen (Descartes research – rationality, boundaries, and the peripheral). Iwanami Shoten, Tokyo

The Biomimicry Institute. http://biomimicryinstitute.org

Tsukimoto A (1996) Girugameshu jojishi (Epic of Gilgamesh). Iwanami Shoten, Tokyo

Umehara T (1995) Mori no shisou ga jinrui wo sukuu (A philosophy of the forest will save humanity). Shogakkan Library, Tokyo

Umehara T, Yasuda Y (eds) (1995) Bunmei to kankyou (Civilization and environment), vols 3, 9.
    Asakura Shoten, Tokyo
Wilson EO (1993) The diversity of life. W. W. Norton & Company, New York
Yamaori T, Nakanishi S (eds) (1995) Bunmei to kankyou (Civilization and environment), vol 13.
    Asakura Shoten, Tokyo
Yasuda Y (2007) Kankyou koukougaku kotohajime (Initiating environmental archeology).
    Yosensha, Tokyo

Chapter 11
The Japanese Industrial Revolution Which Did Not Part with Nature

Abstract As the industrial revolution, with its principle of humanity's separation from nature, was sweeping across the world, the Japanese kept their view of nature and, based on this, in the Edo Period (1603–1868) completed what has been called the industrious revolution. It was the fusion of this view of nature with the industriousness nurtured in Edo, which gave birth to the notion of *iki*. The essence of *iki* is unity with nature; pursuing a life of enjoyment with nature as the most fundamental principle. Here, there is a spiritual freedom and a non-competitive society in which losers are not created. By knowing how much is enough in everyday life, people felt related to all things and did not cause environmental overshoot. The concept of *mitate*, a form of metaphoric enjoyment (best known, perhaps, from Japanese temple gardens in which stones and plants may be metaphoric representations—*mitate*—of the world or cosmos), links to the culture of *wabi and sabi*—the aesthetic sense of Japanese art and culture emphasizing quiet simplicity. Quite possibly, this concept of *iki* helped advance the Japanese industrial revolution. This was an industrial revolution which did not lead to mass-consumption, but in which technologies fueling immaterial or spiritual desires played the main role, thus creating links in everyday life to play and entertainment. The development of technology which incorporates a view of nature means reproducing this concept of *iki* in technology. This also means developing technology which "learns from nature how to be ultra low impact and highly functional", "is simple and easy to understand", "encourages communication and community", and "generates affection". Technology incorporating such elements is, indeed, what we call Nature Technology.

Keywords *iki* • Industrious revolution • Nature technology • The beauty of daily necessities • The Edo Period

E.H. Ishida and R. Furukawa, *Nature Technology: Creating a Fresh Approach to Technology and Lifestyle*, DOI 10.1007/978-4-431-54613-9_11, © Springer Japan 2013

11.1   A Philosophy of Unity with Nature

With the industrial revolution as a major impetus, the philosophy of human dominance over nature—a trait apparently built into our genes—led to the creation of a civilization based on underground resources, which in turn caused the global environmental problems we are facing today. Can we, then, not free ourselves from this philosophy of dominance over nature? Apparently, an answer to this question may be found in Asia. In particular, the distinctive view of nature cultivated in Japan's unique environment may provide new insights. It is said that in the civilizations found in the Eastern part of the Eurasian continent, based as they were on rice farming and sericulture, anthropocentric thinking and the idea of human dominance over nature were weak. Yoshinori Yasuda calls the people who created the Western civilizations "field and dairy farmers" and those of the Eastern civilizations "rice (paddy) farmers and fishermen", the difference being the degree to which water was needed for farming. For rice paddy farming much water is required, and thus forests that can preserve water are needed, which in turn means that the killing of the guardian spirit of the forest (as in the Gilgamesh Epic) would never appear in the cultural narrative of the East. In Japan, in particular, the introduction of rice paddy farming came as late as a mere 2000 years ago (the oldest examples of such farming, possibly, date back as far as 14,000 BCE, to the Chinese Chang Jiang culture). As a result of this, people in Japan did not see themselves as special creatures, but rather as being of the same nature as all other living things. Humans, while struggling with other living things in a harsh environment, would still aim for unity with nature. Thus it has been said that the principles of hunter-gatherer societies were still strongly present. Then, in the sixth century, Mahayana Buddhism was introduced to Japan, and as the philosophies of Saicho (founder of Tendai Buddhism), and Kukai (founder of Shingon Buddhism) merged toward the end of the Heian Period (794–1185), a distinctly Japanese form of Buddhism took shape. In the Kamakura Period (1185–1333), the imported Buddhism was completely Japanized and the major Buddhist philosophies had as their common strand the Tendai teachings of innate enlightenment (in Japanese *Tendai Hongakuron*), a philosophy which had at its core the idea that all living things—mountains and rivers, grasses and trees, and all the land, are imbued with the Buddhist spirit (In Japanese: *Sansen-souboku-kokudo-shitu-kai-joubutsu*). Here, we obviously do not find the idea of human, or even animal, supremacy; rather, all living things including humans are seen to be part of the same cycle of life. This kind of thinking is the same as found in the indigenous beliefs of the Jomon Period (from around 10,000 BCE to 400 BCE) and beyond; namely that all living things are equal and share the same spirit of life. Buddhism in Japan was transformed by such strands of indigenous thought and was re-moulded into a uniquely Japanese form of Buddhism. During this process, the so-called manifestation theory holding that (indigenous) Shinto gods were actually manifestations of Buddha, and the syncretism of Shinto and Buddhism progressed in a multitude of ways.

It appears likely that the unique natural environment found on the Japanese archipelago helped make such views of nature even more distinct. Japan is a country in which the ground often trembles. The geographical area of Japan is only 0.25 % of global land area, but the country has 7.1 % of the world's volcanoes, and 20.5 % of earthquakes above magnitude 6. Thus, the very ground beneath our feet trembles quite frequently and dramatically; in such a situation there is little a human being can do and little to hold on to. Also, the greenery in Japan almost violently imposes itself on human beings. The small mountainous hills found near Japanese villages were used in many intelligent ways. Fallen leaves or forest undergrowth would be used as fertilizer, and wood would be used to supply charcoal or firewood. Just in front of the house, there would be gardens; further out, fields; yet further out, the small mountainous hills, and behind these, the untouched, deep mountains—the latter a place worshipped and treasured as the residence of the gods. Even further away, these deep mountains would connect to the high peaks, which were like ladders to the heavens. The small mountainous hills served as a buffer zone between the deep mountains and populated villages—a buffer ensuring that the wildness of the deep mountains did not impose itself too strongly on village inhabitants. Actually, many efforts were constantly made to protect the mountainous hills from the threatening wilderness of the deep mountains. Grasses and weeds would have to be pulled out again and again—this is the nature of Japanese greenery and the power behind the continued existence of forests on more than 60 % of the Japanese archipelago (By comparison, the large scale cultivation taking place around the twelfth century in Europe meant that 90 % of wooded land disappeared in England, 70 % in Germany, and 90 % in Switzerland). This is how strong and robust the greenery is in Japan. In Europe or the United States, turf would often be laid in church grounds or other sacred places. But in Japan, the green was excluded, thus creating another kind of sanctuary, where—both in the case of Shinto shrines and Buddhist temples—large, round stones would cover the ground. The green in Japan is this powerful—quite different from the idyllic image espoused in, for example, Europe. The same was the case with rivers. When heavy rain fell, the large rivers—where the dragon god was said to reside—would swell with anger and swallow up all in its surroundings. To Japanese, nature is a thing to be held in awe, but at the same time, we know it is this power which gives us life. The ideas held of nature in Japan are far removed from any thinking of human dominance.

If we can say that technology, based on the idea of human dominance over nature, has triggered the present environmental problems, then the approach needed to solve this problem must entail a new awareness of nature built into technology as well as a rethinking of the relationship between nature and human beings. It is our belief that the Japanese, with their uniquely deep view of nature, has the potential to provide solutions to this conundrum.

The historian Arnold J. Toynbee pointed to the fact that between the sixteenth and twentieth century all major civilizations surrendered completely to science and technology, and while the world was conquered by a materialistic paradigm, a great spiritual vacuum was created in the unifying principles of the modern world. What we must do today is not to reject technology, but to create a new kind of technology which helps people live wholesome, spiritually fulfilling lives in a world of environmental constraints. This is also a technology which encourages immaterial, spiritual desires, thus enabling a shift from a world of insatiable material desires based on a philosophy of human dominance over nature to one in which unity with nature is pursued and the fulfillment of our desires is realized in our inner world.

We believe that this Japanese view of nature may become a force for the creation of new forms of technology, leading to a new global standard for technology originating in Japan, which supports a civilization of life and enables new lifestyles.

The movement called Deep Ecology advocated by the brilliant philosopher Arne Naess (1912–2009), who at the age of 27 became professor at Oslo University, was not only a philosophy, but had great influence on the environmental movement. What he meant with "deep" was a questioning of our most basic assumptions, like for example "what does it mean to be prosperous?" What, then, does richness or prosperity mean to the movement of Deep Ecology? The answer is to achieve self-realization. The "self" here does not only indicate the individual self such as "me" or "you". The self includes not only your family, friends, or humanity as a whole, but also is a product of our unity with all other living things. Being able to sense this unity is what Arne Naess called self-realization. Also, in Deep Ecology, one of the most basic tenets is that "all nature, in principle, has the right to live and flourish". This means protecting the diversity of life and placing intrinsic value on this diversity. In this way, the idea of humanity's unity with all living things takes shape. The imaginary enemy of Deep Ecology is the Cartesian mind/body dualism, and a modernity in which humans and nature are regarded to exist in opposition to each other. Naess said that "we may well say that Deep Ecology includes religious or fundamentally intuitive aspects". Ryo Yoshimoto writes the following about Deep Ecology: "Deep Ecology argues that 'religion' or 'intuition', which lost their authority due to and were disdained by modern science, are the elements of life with true value. From the viewpoint of modernity, what Naess called 'intuition' merely means making decisions based on incomplete information, but behind the emergence of such thinking (as that of Naess) we find a soul-searching in relation to modernity which slighted intuition and worshipped rationality". The argument is that all things are of equal value, that we humans are imbued with life by nature, and that we must, therefore, be strongly aware of the relatedness of all living things. This view of nature, however, is nothing new to the Japanese.

In Japan, there are unique concepts of the way to understand nature. One such concept is expressed in the Shinto idea of myriads of gods residing in all elements of nature (In Japanese *Yaoyorozu no Kami*), another in the Buddhist idea, mentioned above, that all living things—mountains and rivers, grasses and trees, and all the land, are imbued with the Buddhist spirit (In Japanese: *Sansen-souboku-kokudo-shitu-kai-joubutsu*). These suggest that all things are of equal value and all are of Buddha nature or possess the divine spirit. An actual culture of living in which such veneration for nature was held while people enjoyed life—and in which people aimed for unity with nature living vigorously but with a minimal impact on the environment—was, we believe, found in the *iki* culture of the Eco Period.

The first Shogun order to close Japan to foreigners was issued in 1633 by Iemitsu Tokugawa. According to history books, Japan's isolation from the outside world was completed with the fifth order to close the country issued in 1639, but in reality things were not that straightforward. After this date, the policy of isolationism was enforced through incidents involving foreign ships (the Dutch Breskens and the English Return), but trade was maintained with Holland, China and Korea, and it was only around the year 1800 that all products needed could be procured domestically, finally enabling the completion of Japan's isolation. To procure all that was needed domestically for a population of some 30 million people was an immense task, and the success in doing so has, by economist Akira Hayami, been dubbed the industrious revolution. This was a revolution in productivity which, in order to save on capital investment, used a concentrated input of labor to raise the productivity of the land. This stands in contrast to the British industrial revolution, which involved the heavy investment of capital in industry and the massive introduction of machines in order to reduce labor input, in the process creating the concept of mass production.

The fact that people in Edo Japan did not waste, or make light of, the limited resources available was a major factor in the success of the industrious revolution, and as a result, enabled the reduction of capital input.

To make this possible, a path of very severe labor was chosen in which even the use of horses and oxen was reduced and substituted by human power. Why, then, did such an approach not lead to a culture of human despair caused by the harsh and unhealthy conditions? We think this is because of the Japanese view of nature which holds that all elements of nature possess a divine spirit. The ability of the Japanese to empathize with all things, and their unwavering belief that all things have emotions, made it possible to succeed with an industrious revolution in which humans willingly took over the work of oxen, and at the same time helped nurture an astonishing aesthetic sense, as well shall see later.

The Edo Period was an era, unique in world history, which saw no war for 265 years (1603–1868), allowing culture to flourish and technologies to develop that were not for the purpose of waging war, but genuinely for the benefit of common people. Edo was thus a rare example of an era which was truly shaped by and among common people.

Many people from local regions gathered in Edo (the present Tokyo) and created a huge metropolis, In Edo, the number of samurai warriors and commoners was almost identical, but the land area provided to commoners was only 15 % of the total. In the very cramped lives of ordinary people there was almost no privacy. In order to make it possible for 800 people to live within one hectare, daily life came to revolve around long partitioned row-houses (In Japanese *nagaya*), which also became media for communication. If, for example, a man in the neighbourhood had to go yet another day without work and towards early evening shouted, "I am starving, would someone give me a bit to eat!", someone in the row-house would congenially reply, "The poor chap fell out of work again today, what a pity" and then prepare a meal for the man. We imagine this is the kind of community found in downtown Edo. People would not hoard money for the future and enjoyed Kabuki performances, *Rakugo* storytelling and the numerous show booths and tents found in the city with unusual items, animals or funny performances. Maybe part of the reason was the frequent fires, part of the reason the narrowness of living quarters, but in any case, people were not carried away by material urges, and thoroughly enjoyed the entertainment of the day. In the show booths, when mechanical archer dolls hit or missed their target, the audience would roar with excitement, and if it was a miss, would nervously expect the doll to hit target in the next round. Norinaga Motoori, an Edo scholar of classical Japanese studies, praised the Tale of Genji in the following way: "This story is excellent for understanding the pathos of things—*mono no aware*" and described the concept of *aware* as "the deep feeling of mindfulness that is gained from all that you see, hear and take part in". *Mono no aware* thus means reflecting yourself in other (things) and through this objectification expressing yourself. All things in nature are regarded to be of equal value as humans, and through the empathy felt for other existences, it became possible to express your own innermost feelings and sensitivities. Robots, which in the West often are seen to be in conflict with human beings, are given nicknames in Japan; industrial robots are polished and the operator may even talk to them; the ornamental dolls put on display for Girl's Day are called "o-hina-sama" thus personified using both a prefix (o) and suffix (sama) indicating respect; people keep their cars shiny and for the New Year's celebration decorate them with ceremonial ornaments—and no one finds it odd for cars to drive round town with such adornments. These examples show how people not only empathize with things, but also the way in which the belief that all things are imbued with a divine spirit permeates Japanese thinking.

As this shows, "things" to the Japanese do not mean the same as when a Westerner talks about physical "things". Things in the Japanese perspective are like a medium between nature and human beings. This is quite different from, for example, the Christian perspective in which God is a kind of medium connecting people to each other. In Western thinking, material things are of the Earth whereas spiritual things belong to the Heavens, but to the Japanese, physical and spiritual "things" are of the same nature and both exist on this Earth.

The divine dwells in all things—even robots, a pair of chopsticks, or withered leaves fluttering to the ground, and we humans can spiritually identify ourselves with this divinity. This is the reason why, in Japanese manufacturing, an approach of wholehearted involvement and meticulous dedication to the process is favored. Things that have been manufactured in this manner are not easily thrown away, but rather encourage people to use them for lifetime. Such are the origins of the making of things in Japan.

From such a lifestyle revolution aiming for harmony with nature, many technologies were born in the Edo Period, in the process shaping a particularly Japanese industrial revolution.

The sublime movements of mechanical dolls found in show tents in the city are one example of how, in the peaceful environment of Edo society, technology was popularized. Ikkansai Kunitomo, an inventor and gun smith in late Edo, learned the mechanism of air guns through a gift presented by the Dutch government, and applied this in the making of a lamp with long durability for popular use called "the never-ending lamp". Western style mechanical clocks introduced into Japan were not of much use in the farming practices in Japan, and were reinvented as a Japanese clock using the so-called "indefinite time method", a way of measuring time by equally dividing daytime and nighttime, as determined by the times at which dawn and dusk occur. Through the dynamic curiosity and inventiveness of the Japanese, advanced technologies imported from abroad were, one after another, transformed into items of practical use to people in their everyday lives. Technology was not the possession of a limited group of experts, but was brought to life in the everyday. This is, indeed, the essence of a cultural attitude found in Japan—the "combination of the Japanese spirit with Western knowledge"—*wakonyosai*.

Why, then, despite a widespread popularization of technology occurring earlier than in many other countries, did Japan not create a wasteful society of mass production and mass consumption? In the Edo Period, the highly sophisticated view of nature held by the Japanese merged with technology, which then became available even to common people—a process of technology development and diffusion quite rare in the world. Here, skills unconsciously discovered through play and skills used to entertain other people were regarded to be valuable, and a spiritual culture was naturally cultivated in which entertainment fueling spiritual rather than material desires was seen as the preferred path to a life of enjoyment. This is where a "technology wrapped in culture" was born, and perhaps we can say that it led to an industrial revolution which incorporated a view of nature.

These currents of thinking still flow strongly in the hearts of the Japanese. At least, in the aftermath of the Great East Japan Earthquake, as the shiny makeup of financial capitalism seemed to peel off, such a philosophy was very much alive. Before the makeup is again put back on, we need to recapture the original beauty of the face of society.

Matthew C. Perry, whose arrival to Japan on the "black ships" eventually led to a reopening of the country, was amazed at the fine craftsmanship he found in Japan. The English poet Edwin Arnold eulogized Japan, writing that both its landscape and people must have descended from the Heavens. Griffiths, who came to Japan as a teacher, stated that the main work of this country for the last two centuries had been entertainment, and, noting that there seemed to be no boundaries between adults and children, apparently found Japanese adults to be truly admirable. Aime Humbert, head of a Swiss delegation to Japan, to his satisfaction found that a simple life was a life of leeway and freedom, and the French painter Felix Regamey noted that the Japanese smile was the basis for all etiquette. Henry C. J. Heuskens, who was interpreter for the American consul Harris, wrote that introducing Western civilization to the country of Japan might well end up being an evil undertaking. Edo (the present Tokyo) had the largest population of any city at the time, and also enjoyed the highest rates of literacy and levels of hygiene in the world. It is this historical accumulation of skills and culture which, in the long term, enabled Japan to become the scientific and technological nation it is today.

11.2   *Iki*: The Spirit of Edo

While enjoying entertainment and developing world class skills and technologies—as epitomized in the refinement of mechanical devices or the selective breeding of carps—no industrial revolution leading to mass production and mass consumption occurred in Japan. The culture of *iki* refined in the Edo Period was the natural response of the Japanese to the challenge of how to create a sustainable society. Shuzou Kuki, a student of Heidegger, defined *iki* as, "a sophisticated, vibrant form of coquettishness". People would not boast, but subtly express themselves; they would not pay much regard to money and would not spend large amounts on things with a physical shape (architecture), but rather use the money they had on the ephemeral (*Rakugo* storytelling and other performances, *Kabuki* Theatre, *Sumo* etc.); they would first of all try to ensure that their friends were having a good time, and they would experience a sense of quiet satisfaction in the simple beauty of a single flower or the freshness of a light breeze—a sensitivity expressed in Japanese as *wabi-sabi*. In the culture of *iki*, all things including humans were seen to be of equal value and people would enjoy their everyday while paying respect to nature, all of which was thought to be of divine or Buddha-nature. This was a culture which nature had helped give birth to, and here we find a philosophy of living which, we believe, is essential to create an altruistic society.

There are four important elements in the culture of *iki*. The first is living in harmony with nature; enjoying life to the full without placing a large burden on the Earth. It is a culture in which satisfaction is derived not from the purchase of things, but rather from the enjoyment of various forms of entertainment—watching performances at the street tents or going to a *Kabuki* show or *Rakugo* storyteller's performance with friends or neighbors. Secondly, the culture of *iki* was one which aimed not to create losers in society. Mainstream elements of culture and society were not reserved for the rich and the strong; special attention was also paid to people who would normally be seen as losers; indeed, it was a culture in which such people had the opportunity to play key roles. It was, thus, a society in which "the common Joe", even if he were out of work, could live unashamedly. The third element of *iki* culture was a sense of moderation—of knowing how much was enough. People would not seek to gain more than was needed in daily life, and all things would be used to the full in respect for life and nature. As best expressed with the phrase *mottainai*—all things are too precious to waste—it was believed that the gods resided in all and everything and gratitude was felt even for a single grain of rice. The fourth element of the *iki* culture involved the metaphoric enjoyment of nature. We could say this is the same as the *wabi-sabi* culture in which the greatest satisfaction is found in the simple and the quiet. Even if people had money to buy a mountain, there was no need to do so; by arranging sand in your small enclosed garden and pretending this was Mount Fuji, you had achieved the same sense of satisfaction in a metaphoric manner. This was called the culture of *mitate*, and is also found in, for example, the traditional tea house which is a representation of cosmos.

To summarize, the essence of the *iki* culture was to aim for harmony with nature; to live and enjoy everyday life with nature as the foundation. There was a mental and spiritual freedom, and since the principle of competition was not strongly present, losers were not created. By living a life of moderation (the spirit of *mottainai*), overshoot did not occur in any of the relations humans had with their surroundings. And, finally, a cosmological worldview in which the metaphoric enjoyment of nature was practiced meant that even everyday life was connected to the *wabi-sabi* culture of discovering beauty in the simple and the quiet. *Iki*, to us, appears to be a truly unique form of culture. Both the unemployed neighbor's demand for a bit of food for supper and the entertainment found in the mechanical archer dolls are part of the concept of *iki*.

11.3   The Japanese Industrial Revolution

The main industry of Edo Japan was, obviously, agriculture. Things needed in daily life, including for example energy (firewood, charcoal, wooden torches, vegetable oils, wax), were procured and made from the land. Also, since land taxes were paid in rice, society could not function without a market. Farmers did not live isolated lives of self-sufficiency tied to the land, but apparently enjoyed a great range of freedoms, and in the middle period of Edo, commerce as wells as a coin-based currency were indispensable.

The Japanese population at the time was some 32–33 million—about a fourth of what it is today—and was the fifth largest in the world, surpassing both the United States and England. This large population was maintained in a country closed to the outside world, with agriculture as its main industry and all things needed manufactured domestically. There was not much leeway when it came to the ability of the environment to provide for all this, and thus, regulations on the use of forests were introduced and jointly administered by bureaucrats and common people, and wooden material was thoroughly reused and recycled. The town of Edo (Tokyo) was one of the largest cities in the world; it was extremely densely populated and the number of inhabitants may have reached some 1.3 million at the end of the Edo Period. Enormous flows of food and fuel entered Edo, and human excretions and waste also increased. In contrast to the situation in European cities, though, human excrements were collected and used as fertilizer, and it was strictly forbidden to dispose of trash and waste on streets, empty land, or in the moats and rivers. As a result of this, Edo was kept so clean that it astonished all visitors coming from abroad.

Edward S. Morse, an American zoologist and orientalist who discovered the Omori shell mound in the town of Edo, was amazed at the level of culture found in Japan. The literacy rate of people in Edo was 98 % (*Hiragana*, at the beginning of the nineteenth century), and both people's interest in culture and their level of education were the highest in the world. Towards the end of the Edo Period, the school attendance of children had reached more than 80 %. Not many things were needed in daily life; living was simple and prices low, and if only you had learned a craft or mastered handiwork and worked diligently, you would have enough even to enjoy a cup of *sake* before bedtime. There was no need to be bothered by much else in daily life, and, apparently, life in the town was a micro-cosmos on its own, there was always plenty of laughter, and even when a person had to leave this world, this was calmly accepted much in the same way as when the withered flowers of a cherry tree flutter to the ground.

The average monthly rent of a row-house in Edo was 400 *mon*, the equivalent of some 6,000 yen (app. USD 60) today. It is said that a craftsman could earn twice that amount in a single day, and even a walking vegetable peddler, if working hard, could earn some 4–500 *mon* a day. In the beginning of the nineteenth century, when Edo's city culture was at its peak, living costs for the middle class were around two *ryo* per month, the equivalent of some 160,000 yen (app. USD 1,600) today. If a family of four earned two *ryo* and two *bun* (some 120,000 yen/USD1200), there would be money enough left to enjoy a nightcap every day. If townspeople worked 10–15 days a month, they would earn enough for a living, and apparently many chose to do just that rather than toiling and moiling.

The quarters where common people lived were extremely densely populated, communication was lively and thick, people virtually lived "nose-to-nose", and their shoulders would bump into each other when people passed through the narrow alleys. The people in Edo did not at all complain about his situation, though, but rather accepted it as the natural state of affairs. Daily life and neighborhood relations were held up by common people's ability to endure and be patient, and in order to live in harmony, people most likely were cautious to be considerate of others in their daily undertakings.

Tadataka Inou, an Edo merchant, learned how to be a surveyor thanks to his strong curiosity. The same was the case with Takakazu Seki, a scholar of Japanese mathematics, said to be the most advanced in the world at the time. Many such mathematics experts honed their skills by posing mathematical questions for each other to solve. Doing so, they were not even thinking about making money from this for a living. Interestingly, the results of such jovial contests were hung up in shrine or temple grounds, much like the wooden prayer boards one can find on such premises also today. These efforts later came to be of much use in land surveying which requires complex calculations. The Japanese cultural pioneer, Yukichi Fukuzawa, in his autobiography wrote the following about the characteristics of such scholars in Osaka: "These scholars, however poor and troubled, however poorly clad and fed, at first sight look to be nothing but pitiful scholars, but the active and elevated mental skills you find in these people are so strong that they would not hesitate looking down on even an aristocrat at court…the more difficult the question at hand, the larger the joy for these scholars; their mental circumstance is such that they are able to find joy in hardship, or even turn hardship into joy". Pursuing and perfecting a "way" or school without any particular purpose—maybe that was what became one of Japan's strengths.

The hand-operated electric generator or the pedometer invented by Gennai Hiraga, a scholar active in mid-Edo, were also products of curiousity. Shouzan Sakuma, Ryoutaku Maeno, Konyou Aoki, Michitaka Kume, Benkichi Oono….all these scholars and inventors appear to have been ingenious at play and amusement. And, as a result thereof, *Ukiyoe*, for example, created with more than ten woodblock prints made one after the other (with a precision of 1/10 of a millimeter to fit each print on the previous one. Considering that the required precision for the 1970 masking techniques in integrated circuits was 1/100 of a millimeter, this was an amazingly high precision), were sold for the equivalent of a weekly magazine today. If we can say that it is the accumulation of knowledge which creates culture and the accumulation of technology which enables civilization, then this is, indeed, a splendid example of one of the outcomes of a fusion of culture and civilization in Edo society.

With the view of nature which the Japanese naturally held as a whetstone, it is, we believe, possible to sharpen the inner senses once more. Rather than looking to religion for an answer, the reverence for nature found, for example, when a Japanese naturally puts his hands together and bows in gratitude for nature's blessings is a spirit which, we believe, should even be taught to children today.

The town and era of Edo were a time and place in which people made the greatest possible effort to realize an altruistic society while enjoying entertainment and using all things with the greatest possible care in everyday life. This kind of culture was learned and acquired, built on the foundations of an intimate view of nature never lost and formed by the industrious revolution. For this reason, unless we consciously maintain this culture, it will be lost.

We cannot return to the Edo Period, but it is possible to learn from it. The ways of living discovered through our interviews with nonagenarians are great treasures which, indeed, need continuous maintenance lest we should lose them.

At least, we can say that there was a truly marvelous interplay between the *iki* culture of Edo and the civilization created through the accumulation of numerous technologies. Up through history, technology has mainly been developed to wage wars, or for the benefit of the privileged classes or the aristocracy. Therefore, if we define an industrial revolution as the popularization of technology, Edo Japan had actually spurred on an industrial revolution more than 100 years before Great Britain. According to Kazuyoshi Suzuki from the National Museum of Nature and Science in Tokyo, many things in Edo were available through catalogue sales; for example, the surveying instruments developed by Tadataka Inou could be purchased by common people from a catalogue. It is, however, also worth noting that these technologies did not lead to mass production or mass consumption. What, then, were the set of values upon which these technologies were developed? For example, the mechanical archer dolls mentioned above were mechanical contraptions that picked up four arrows, one after another, aiming at a target. The dolls would display amazingly delicate movements using only a simple spring coil made from whalebone. In Edo society, such technologies were not applied in manufacturing, but were enjoyed in entertainment and various performances. The archer dolls would draw their bows as a performance, and the audience would be fascinated by each and every of the doll's graceful movements, hoping anxiously for it to hit target. This is Edo technology, created by the *iki* culture—a way of living based on four elements; the avoidance of wasteful behavior; aiming for harmony with your inner self; enjoying entertainment; and the lack of a pursuit of mass production. In a Japan where there had long been no wars, the principle of the air gun was applied to create a "never-ending lamp", and in the same way, the sophisticated technologies of Edo were used to create many things—such as Gennai Hiraga's generator of static electricity, the Japanese clock applying the indefinite time method, etc.—that were part of a spiritual culture encouraging entertainment rather than of an industrial culture fueling ever growing material desires.

There was a passion about things in Edo to such an extent that people visiting from abroad were amazed that the Japanese were using what appeared to be art objects for daily purposes. This is what came to be called the "beauty of daily necessities"—in Japanese *you-no-bi*. This is the very world of *iki* and is also what created an affection for things; it was the motivation behind the careful and long term of use of both things and technology. The *iki* culture of Edo, which never forgot to take nature into account, and the Edo technologies which this culture helped create show us an excellent of example of the interaction and mutual inspiration of culture and civilization.

Let us take one more look at the differences between the British industrial revolution and the Edo ditto occurring more than 100 years earlier. In England, the industrial revolution made it possible to manufacture goods in much larger volumes and to transport both people and goods much further than earlier, thus leading to major changes in the social structure. What was the situation like in Japan? As we have seen, many sophisticated and precise technologies were born, but most of them did not leave the sphere of entertainment or play. Does this then indicate that the British industrial revolution was innovative whereas the Japanese was somewhat behind? We do not believe that is the case. Maybe the Japanese were simply not particularly interested in using technology to travel long distance or to produce large volumes of food or goods. Of course, being materially well enough off to live a decent life was a precondition also in Japan, but as long as people could maintain a proper everyday life, they did not yearn for more money, and, it appears, placed the highest preference on how to enjoy and have fun. This is, indeed, the way of manufacturing enabled by the industrious revolution. Against this background, technology often took the shape of skills developed unawares in a flow of cultural entertainment, and value was placed on skills that made people enjoy life. A spiritual culture with nature as its mentor was nurtured in which people enjoyed everyday life, and which fanned people's immaterial rather than material desires.

Textiles, lacquer ware, craftsmanship…all forms of technology were perfected to the level of art, and it is no exaggeration to say that these craftsman skills are what came to form the backbone of Japan's future technological development. This is how mature Edo society and the technologies nurtured therein were. Play became a mainstream activity in society and contributed not to a system of mass production, but to the development of a highly sophisticated culture.

What many foreigners coming to Japan were all amazed at was the cleanliness of the country, and the fact that even daily items used by poorer people in society looked more like objects of art. Muneyoshi Yanagi, a philosopher and scholar of art, described this in the following way. Items of daily life are anonymous, and for that very reason they are free. From this freedom is born truly useful and beautiful items; because these are anonymous they are also free of greed, because fame is not pursued they are unselfish, and this was the very soul (the pursuit of the beauty of everyday necessities) of the master.

In our view, this is a fine description of the essence of the Japanese spirit of play and enjoyment. Here, play and enjoyment are of a fundamentally different nature than the pursuit of instant pleasure. Heita Kawakatsu, author and governor of Shizuoka Prefecture, in his book "Fukoku Yuutokuron" (A rich nation with virtues) wrote that the worldview of Tokugawa Japan (The Edo Period) was not one based on the contrast between "war and peace", as in the West, but rather on the difference between "the elegant and the unrefined", that is, focus was on the difference between "civilization and barbarism". Here lies, perhaps, the origins of the Japanese spirit which helped elevate "play" to the level of "schools" or "ways". In Shintoism, to play is called "to let the flowers of joy bloom"—a wonderful expression and a necessary spirit to be able to experience mental sophistication even when living a humble life.

11.4   The Four Elements of Technology We Can Learn from *iki*

The concept of *iki*, which we have tried to describe from different angles in this chapter, is a basic concept of Nature Technology. The industrial revolution triggered the development of modern technology, but the success of this enterprise was gained only by removing humans and their technologies from nature. This philosophy of human dominance over nature became a key principle of modern technology and civilization, and led to an ever great ever burden being placed on the Earth. Reintroducing a sense of nature into technology—that is the reshaping of technology much needed today. In Japan, there was no removal from or parting with nature. Actually, it may be more correct to say that in a country such as Japan where the ground shakes, the rivers flow over at major rainfalls, and the greenery imposes itself on people, it was not possible to remove ourselves from nature. This reality was what helped create the culture of *iki* in Edo and a Japanese style industrial revolution which was entirely different from the one originating in Great Britain.

As described in the above, the culture of *iki*, which has not lost a proper view of nature, consists of four elements. They are, "enjoying life with nature as a foundation", "the metaphoric concept which includes a cosmological dimension", "avoiding to create losers", and "moderation—knowing how much is enough".

Transferring these ideas onto concepts of technology, we arrive at the following four elements. Technology that "realizes high function/ultra-low environmental impact with nature as a point of departure", "is simple and easy to understand", "encourages communication and community", and "inspires attachment and affection". A technology which incorporates such elements is, indeed, what we call Nature Technology.

11.4.1   The First Element: Technology Which Realizes High Function/Ultra-low Environmental Impact with Nature as a Point of Departure

Nature drives perfect cycles with a minimal input of energy. Nature Technology looks at such mechanisms and systems through the lens of science and redesigns technology so as to incorporate them.

(At our present level of scientific development, merely mimicking nature may often lead to a situation in which huge inputs of resources and energy are required. Therefore, it is imperative that we, upon gaining an understanding of nature's mechanisms and systems, look at these through the filter of sustainability and reconstruct—or redesign—them as new forms of technology).

11.4.2 The Second Element: Technology Which Is Simple and Easy to Understand (Clear)

The *iki* notion of metaphoric concepts incorporating even a cosmological dimension translates into the idea of simple and easily understood (clear) technology. This is a technology that is approachable and understandable to anyone. There are no "black boxes", and since it is a technology that both adults and children can easily understand, it also becomes possible for anyone to participate in the technology. Of course, the face of the engineer behind the technology must also be visible. It must be a technology which allows us, the users, to confirm important aspects of safety or secure use, and which thus allows for a dialogue on an even footing between the user and the engineer.

11.4.3 The Third Element: Technology Which Encourages Communication and Community

This is an element of technology important in lifestyle design. A technology around which people might gather almost as if it they were enjoying each other's company at a camp fire, and which allows people to connect and initiate conversations. Unfortunately, modern technologies—be it mobile phones or the internet—lack in aspects such as human warmth or sympathy, and have thus evolved along a path far removed from the original purpose of communication. The communication that naturally occurs when you go to the neighbor to borrow *miso* or soy sauce, or the conversation that naturally starts around a camp fire even amongst strangers—we need technology that can play a role similar to that of *miso* or soy sauce or camp fires in the community.

11.4.4 The Fourth Element: Technology Which Inspires Attachment and Affection

The user does not get tired of a product made with robust, quality technology, but rather wants to continue repairing it for long term usage. When such a product is used over time, a feeling of affection arises and the product becomes one for which strong attachment is felt. This is the reason why foreigners visiting Japan in the Edo or Meiji Period (1868–1912) were astonished at the almost artistic level of everyday items and often said that the Japanese were surrounded by objects of art in their everyday lives. This situation existed thanks to the spirit of affection deriving from people's belief that the divine dwelt in all things and that even everyday items should therefore be used with great care over long time.

If the black boxes in technology disappear, and it is simple and easy for anyone to understand, technology will become a closer and more intimate existence in people's lives. Such technology encourages communication and community, which over time leads to a sense of attachment and affection. This is the kind of technology we are in need of today.

One should not believe the already outdated myth that the more black boxes included in technological development, or the more complicated the technology, the stronger will be a company's ability to differentiate itself and compete with other companies. What is needed today is technology than can be understood by anyone, and the premise upon which this is created are visions of wholesome, fulfilling lifestyles even under severe environmental constraints. Differentiation in the future will derive from the ability to manufacture products on the basis of an ability to envision and make proposals for new lifestyles.

Bibliography

Carter VG, Dale T (1974) Topsoil and civilization. University of Oklahoma Press, Norman
Drengson A, Yuichi I, Yuichi I (supervisor of translation) (2001) Diipu ekorojii (Deep Ecology). Showado, Tokyo
Esty DS, Winston AS (2006) Green to gold. Yale University Press, London
Fukuzawa Y (2009a) Fukuou jiden (Autobiography of Yukichi Fukuzawa). Keio University Press, Tokyo
Fukuzawa Y (2009b) Bunmeiron no gairyaku (An outline of civilization studies). Keio University Press, Tokyo
Hawken P, Lovins AB, Lovins LH (2008) Natural capitalism, 1st edn. Back Bay Books, Tokyo
Hayami A (2001a) Rekishi jinkougaku de mita nihon (Japan seen from the angle of historical demographics). Bungeishunju, Tokyo
Hayami A (2001b) Rekishi no naka no Edo jidai (The Edo period in a historical perspective). Fujiwara Shoten, Tokyo
Ishida H (2009) Channeling the forces of nature. Tohoku University Press, Miyagi
Ishida H (2010) Atarashii kurashi to tekunorojii wo kangaeru iinkai (Committee on New Ways of Living and Technology) Chikyuu ga oshieru kiseki no gijutsu (Miraculous technology the Earth teaches us). Shoudensha, Tokyo
Ishi H, Yasuda Y, Yuasa T (2001) Kankyo to bunmei no sekaishi (A world history of the environment and civilization). Yosensha, Tokyo
Itoh S, Yasuda Y (eds) (1996) Bunmei to Kankyo Dai 2 Kan (Civilization and environment), vol 2. Asakura Shoten, Tokyo
Kaneko T (2005) Edo jinbutsu kagakushi (A science history of people in Edo). Chukou Shinsho, Tokyo
Kawakatsu H (2000) Kokufu Yuutokuron (A rich nation with virtues). Chuou Bunko, Tokyo
Kawakatsu H (2006a) Bi no kuni Nihon wo tsukuru (Creating a beautiful Japan). Nikkei Business Bunko, Tokyo

Kawakatsu H (2006b) Bunkaryoku – Nihon no sokojikara (Cultural power – the fundamental strength of Japan). Wedge, Tokyo

Kawakatsu H, Yasuda Y (2003) Teki wo tsukuru bunmei – wa wo nasu bunmei (Civilizations that create enemies – civilization that create harmony). PHP Publishing, Tokyo

Kitahara S (2003) Hyakumantoshi Edo no seikatsu (Life in Edo – a city of a million people). Kadokawa Sensho, Tokyo

Kosuge H (2008) Edo na ikikata (An Edo way of living). Tokuma Bunko, Tokyo

Kuki S (1998) Iki no kouzou (The structure of iki). Iwanami Bunko, Tokyo

Morse ES, Ishikawa K (transl.) (1990) Nihon sono hi sono hi (Japan day by day). Heibonsha, Tokyo

Naess A, Saito N, Hiraki T (transl.) (1997) Diipu ekorojii to ha nanika (What is deep ecology?). Bunka Shobo Hakubunsha, Tokyo

Oppenheimer S (2004) Out of eden: the peopling of the world. Robinson Publishing, London

Shiba R (2007) Nihonjin he no yuigon (My last words to the Japanese). Asahi Bunko, Tokyo

Shiba R, Donald K (2007) Nihonjin to nihonbunka (taidan), (The Japanese and Japanese culture (Dialogue)). Nakamatsu Bunko, Tokyo

Suzuki K (2006) Mite tanoshimu Edo no tekunorojii (Edo technology enjoyed with the eye). Suuken Shuppan, Tokyo

Tanaka Y (ed) (2003) Edo no iki (The iki of Edo). Kyuryudo, Tokyo

Toynbee AJ (1985) A study of history, abridgment of volumes I–V. Oxford University Press

Toynbee AJ, Hasegawa M (transl.) (1988) Rekishi no kenkyuu 1–3 kan (A study of history), vols 1–3. Shakaishisosha, Tokyo

Tsuji T (1992) Edo jidai wo kangaeru (Contemplating the Edo Period). Chukou Shinsho, Tokyo

Umehara T (2006) Nihon bunkaron (A study of Japanese culture). Kodansha Gakujutsu Bunko, Tokyo

Umehara T (2007) Umehara Takeshi no jugyou – hotoke ni narou (Lessons with Takeshi Umehara – let us aim for Buddha-hood). Asahi Shimbunsha, Tokyo

Underwood P (1993) The walking people – a native american oral history. Tribe of Two Press

Wade N (2007) Before the dawn: recovering the lost history of our ancestors. Penguin, reprint edn.

Watanabe K (2005) Ikishi yo no omokage (Vestiges of a fading world). Heibonsha, Tokyo

Watanabe K (2007) Naze ima jinruishi ka (Why be concerned about the history of humanity now?). Yosensha, Tokyo

Watanabe K (2008a) Edo to iu genkei (The phantom image that is Edo). Genshobo, Tokyo

Watanabe K (2008b) Nihon kinsei no kigen (The origins of modern Japan). Yosensha, Tokyo

Watanabe S (2010) Sekai ichi no toshi Edo no hanei (The flourishing of Edo – the world's largest city). WAC, Tokyo

Yanagi M (1981) Yanagi Muneyoshi zenshuu 11. Teshigoto no nihon (Complete works of Muneyoshi Yanagi 11. Japan – a country of craftsmanship). Chikuma Shobo, Tokyo

Yasuda Y (1997) Joumon bunmei no kankyou (The environment of the Jomon civilization). Yoshikawa Koubunkan, Tokyo

Yasuda Y (2004) Kikouhendou no bunmeishi (A civilizational history of climate change). NTT Publishing, Tokyo

Yoshimoto R. Diipu ekorojii (Deep Ecology). http://www.ksc.kwansei.ac.jp~kamata/semi/1999/rf1999/ss/deepeco.htm

Chapter 12
The New World Created with Nature Technology

Abstract Technology created with a Nature Technology development system harnessing the amazing powers of nature has its roots in lifestyles, is simple and easy to understand, has an inherent view of nature, encourages communication and provokes affection in the user. Today, on the basis of lifestyles envisioned with backcasting, many examples of such Nature Technology have been or are being created: An air conditioner without a power supply which learns from termites; a waterless bath learning from the mechanism of foam; a wind generator revolving even in a slight breeze learning from the wings of a dragonfly; and a kitchen garden system which, learning from the diversity of microorganisms, needs no pesticides. And, following these examples, many more technologies are being created right now, one after another. These are technologies created from lifestyles based on backcasting and do not involve any "black boxes", neither are they merely substituting one technology for another. They are not technologies intended solely to realize comfort and convenience, but rather ones that allow for humans to live wholesome, fulfilling lives while regaining some of the skills lost through the "outsourcing" that has taken place in the last few centuries. And, they are technologies which may either help change lifestyles directly, or provide opportunities for people to change lifestyles. For such reasons, we can say that these new technologies will enable us to create thrilling and exciting lifestyles that are also wholesome and fulfilling.

Keywords Air conditioner without a power supply • Backcasting lifestyles • Kitchen gardens needing no pesticides • Nature technology • Soil microbial diversity and vitality value • Waterless bath • Wind generator revolving in a slight breeze

12.1 Technology Assuming Responsibility for Lifestyles

In many regions of the world, air conditioners are essential to maintain a comfortable lifestyle. Efforts to increase the efficiency of conventional air conditioners—through the improvement of heat pumps, targeted air conditioning following people's movement in the room, or the introduction of automatic cleaning functions to avoid inefficient operation—are more or less approaching the limits of what is possible. If we can learn to apply the mechanism of the humps on the fins of a humpback whale which help reduce water resistance, or mimic the shape of seeds from the plane tree carried by the wind in order to improve the shape of fans in air conditioners, a further improvement in efficiency may possibly be realized. However, no matter how efficient the device may get, it will still use energy, and there will be a need for resources and energy to manufacture the device. Merely pursuing efficiency improvements will not help change lifestyles; in the below, let us see what happens when we take a backcasting approach.

12.1.1   <Lifestyle Example I. The Wind Determines the Price of Real Estate>

In the past, when it was hot in summer, people would turn on the air conditioner to achieve a comfortable indoor climate. Such air conditioners removed heat from external air to cool the room, but at the same time, the removed heat was wasted into the environment. As more and more people continued wasting such heat in order to keep living space comfortable, the outside temperature kept increasing, and in cities heat islands were created. The temperature of entire towns and cities went up. In order to live comfortably, people had to cool the hot air, and the more energy consumed from this, the higher the temperatures outside…People became unable to open the windows in summer, and we are told they were no longer able to enjoy the sound of chirping insects or birds. They were supposed to live comfortably, but somewhere along the road were forsaken even by nature.

Even now, one can occasionally find houses where an air conditioner is running, but it is not possible to find any people in these houses. There are also many people who keep the windows of the house wide open to enjoy the wind passing through or the faint smells coming from the sea or nearby mountains. The price of land, apparently, varies greatly depending on what kind of winds blow in the area. Houses today are comfortable even with the windows closed; wall, floor and ceiling materials help control indoor temperature and moist content and even absorb smells. People do not spend their time under the air conditioner, but rather enjoy gathering around the dining table frying dried horse mackerel for supper. This dried fish is handmade with pride by the father of the household; dad looks so cool to me these days.

Even under the severe environmental constraints of 2030, the ability to enjoy comfortable living indoors is a value that we cannot give up, but we need not necessarily rely on the air conditioner as a "thing" or "device". The function of an air conditioner is the "action" of controlling room temperature and moisture and in other ways regulate air quality. Thus, it would sufficient if a room had floors, walls, or a ceiling which sensed the temperature and moist in a room and self-regulated without the use of a power supply.

Fig. 12.1 House with an "earth air conditioner" (the floor), (Okinoerabu Island, Amami Islands)

If we knock the door of nature's treasure house, we find a possible solution in the termite hills on the African savannah. Although the temperature swings between 50 °C in the day time and zero degrees at night, the temperature in the termite hill is kept at exactly 30 °C. In the "earth" of which the termite hills are constructed, there are countless small holes of nanometer size which enable the regulation of temperature and moisture inside. Actually, any kind of earth or soil has these kinds of holes (cohesion structure of clay minerals), and by hardening earth without destroying this structure, the air conditioner without a power supply was born. The floor or walls fulfill the function of the air conditioner. Needless to say, the functionality drops some 20–30 % compared to a conventional, mechanical air conditioner. However, a very interesting phenomenon occurred when this kind of air conditioning system was introduced in one of the authors' home. That the material used for this air conditioner is made from earth (nature) is obvious even to children (simple and easy to understand), and thus all in the family including the children would participate in family meetings (communication) to discuss how to make up for the loss in cooling effectiveness, and in summer water would be dispersed inside the house to achieve extra cooling. In winter, members of the family would put on an extra sweater… and as such actions were continuously taken, affection, interestingly, was born and all in the family started taking good care of the technology. As a result, energy consumption in the household dropped by 20–30 %. This is an example not of how to endure worse living conditions, but rather of how Nature Technology can provide new lifestyles of wholesomeness and fulfillment (Figs. 12.1, 12.2).

Fig. 12.2 People will gather in a comfortable house (Okinoerabu Island, Amami Islands)

Nature Technology is a "simple and easy to understand" approach in which we contemplate what constitutes wholesome, fulfilling ways of living under the severe environmental constraints in 2030, then learn the required technologies from "nature" (Biomimetics/Geomimetics), which then encourage "communication" and evoke "affection". Wall material that does not stain easily and is easy to clean is gradually being adopted in kitchens or for the exterior of buildings and was created through a redesign process based on the principles of the shell of a snail, which is always shiny and clean. This is one form of technology which assumes responsibility for a lifestyle.

(For more detail, see "Channeling the Forces of Nature", or "Chikyuu ga Oshieru Kiseki no Gijutsu" (Miraculous Technologies the Earth Teaches us) by this book's authors).

12.2 A Waterless Bath Learning from Foam

After the Great East Japan Earthquake in March 2011, water supply was severed for as long as three weeks (the author's house). This gave one of the authors of this book reason to feel deep gratitude for the blessing of having water and also spurred many other thoughts. During the weeks with no water, it was possible to live on just 2 L of water/person per day for cooking, washing and cleaning the body with a washing cloth. In comparison, a normal toilet uses some 6–10 L of water in one flush, and when we think of bathing, a shower requires some 15–18 L per minute whereas filling the tub takes some 200 300 L of hot water. The number of households in Japan was approximately 48.64 million in 2010 and, although population continues to decrease, is expected to continue to increase until around 2015. After that a gradual decrease is forecasted, but the total number of households will still be around 48.80 million in 2030. The reason behind this is that single occupant households continue to increase (29.5 % of the total in 2005, expected to reach 37.5 % in 2030). If we assume that all households fill one tub everyday (in Japan it is custom to take a hot tub bath every night, using the same water for the entire family), the volume of water consumed just to do this will reach 14.64 million tons/day in 2030. This is equivalent of some 37,500 25 m swimming pools. Furthermore, since it is necessary to heat this water to 40 °C, tremendous amounts of not only water, but also energy will be required. It is obvious that we will not be able to supply such huge amounts of water and energy in 2030 facing severe environmental constraints.

Then, how will we be able to bathe? With a forecasting mentality, the response might be that we should reduce the number of baths taken or make the time for each shower shorter, but are there no other solutions? With such approaches, people would have to endure scarcity and would not be able to enjoy life. If we start from the premise that we must create wholesome, fulfilling lifestyles that are possible even under severe environmental constraints, what kind of bathing could we envision? A new way of bathing derived from an example of lifestyles envisioned with backcasting has come into view.

12.2.1   *<Lifestyle Example II. Enjoying Shared Bathing>*

These days it is hard to believe, but in the past, people had the custom of taking morning showers everyday, and hot tub baths were ready for use 24 h a day in every household. One can only wonder how they were able to secure enough water and energy to enable this. In Japan, the air is moist and people are cleanly, so tub baths are indispensable. These days, a new trend is "community water sharing". People fill the tub at home and invite friends to come and bathe. Sometimes, even friends of friends will come, and the chat after the bath greatly helps expand the host's social circle. There are more and more events organized by members of this water sharing community, and within the community people take turns to teach classes of, for example, sweets making, DIY, or outdoor skills.

If we think with a forecasting mentality, people have to endure worse living conditions, but with a backcasting mentality, other possible bathing options appear on the horizon. As we have touched on earlier, the structure of our desires is such that perceptions of quality of life are irreversible, and on top of this, in a hot and humid country such as Japan, taking a daily bath is a necessity. What about taking a bath every day, but without using water? A bath which cleanses the body, retains heat and is as relaxing as possible. We do not necessarily have to use water for such a bath.

Let us, again, knock the door of nature's treasure house. How do insects manage to live comfortably? Or what about birds? Some of them use sand to bathe, others even bathe in ants…and thinking of pigs or elephants, we see that they enjoy bathing or rolling in mud. Mud bathing, sand bathing, or ant bathing…these may not necessarily appear to attractive options to us humans. Let us try to knock that door again. The chrysalis of a spittle bug (of the Cercopidea family) is enveloped in foam.

The foam protects from ultraviolet radiation, the plentiful air contained in the foam has an insulating function and reduces the effect of external differences in temperature, and thanks to this, even a weak chrysalis can maintain life in a comfortable environment. The fish called Betta splendens (Siamese fighting fish) lays its eggs in foam. The foam has a surface tension which enables it to bond to various things. Why do insects and animals tend to gather in the basin under a waterfall? When the water bubbles break, ultrasonic waves are generated, and, possibly, these bubbles help remove dirt. The crackling sound of bubbles breaking has what is called a 1/f fluctuation, widely found in nature, and may be generating a relaxation effect.

Thinking along such lines, the possibility that bubbles or foam might lead to new ways of bathing starts to appear. By creating bubbles at a temperature of 70 °C, they become able to retain enough heat to warm a person's body. When the bubbles break, ultrasonic waves are released that help remove dirt, and the bubbles should then bond with the dirt removed. If one could take a bath with soft, airy bubbles holding warm air, the sturdiness of the bath tub need not be as high as previously, and it becomes possible to achieve a reduction in the materials used. Being able to bring your portable bath to bathe anywhere you like might also create new ways of enjoying life. Today, you bathe in the bedroom; tomorrow, on the veranda…People may become able to discover new and more attractive lifestyles. What is more, since the bath tub is only filled with bubbles, a much lower level of water tightness than with conventional baths should suffice, which opens up for the possibility, for instance, of bathing in a wheel chair with a special gate attached. This could create new ways for the elderly, or people with disabilities, to enjoy taking bath a bath. For the elderly, in particular, possible health benefits may be realized from taking a bath in which there is no water pressure on the body. If an effective cyclical system is installed, it is more than possible to take a pleasant bubble bath with a mere 4–8 L of hot water. Taking a bath in a tub with no water pressure feels a bit like floating in space, a strange sensation the authors of this book had never experienced before. There are still several issues to solve, such as how to make stable bubbles without the use of surfactants, but this waterless bath undoubtedly has the potential to unleash new lifestyles (Fig. 12.3).

In this way, environmental constraints are not demanding of us to endure living a harsh life; rather, they are positive constraints urging us to discover new lifestyles we had not even contemplated so far.

12.3   Micro Wind Generators Learning from the Wings of a Dragonfly

12.3.1   *<Lifestyle Example III. A Wind Generator Revolving in Even the Slightest Breeze>*

In 2030, it is no longer possible to use oil to generate electricity as in the old days. Due to the Fukushima Daiichi nuclear accident in 2011, the operation of almost all nuclear power plants has also been halted in the developed countries. Most developing

Fig. 12.3 The bubble bath

countries have also given up the use of nuclear power due to the excessive cost of treating and storing nuclear waste, a task that these countries used to rely on the developed nations to take care of. Instead of such energy sources, countries have started to match their wits against each other trying to harness renewable energy sources that may individually be insufficient, but which are abundant as a whole. The use of energy has also changed considerably. Nowadays, small wind generators set up in people's gardens are all the rage with children. Why? Because with these generators you can save energy to use yourself. Kids in the neighborhood enjoyed playing computer games for a while with the energy saved yesterday. They would also like to use the energy saved today for playing, but having heard this morning that the old lady next door is running out of electricity for her hearing aids, the kids are now thinking about donating the saved energy to her. If only the wind would blow a bit stronger tomorrow…the kids love looking at the revolving wind generator—and the blue sky—while thinking such thoughts.

What kind of wind generator would be needed to realize this kind of lifestyle? Here, one can almost hear a child asking, "Mum, can I play computer games?", and the mother answering, "Yes, but it has to be with the electricity you generated yourself"! Maybe the wind generator needed to enable such a lifestyle is one which will revolve even in the slightest breeze, evoking memories of summer days in Japan when small wind bells hanging in front of the house are tinkling in the light summer wind. In this lifestyle, a generator is needed that will be revolving almost all the time—at least more than 20 out the day's 24 h. A wind generator that, in this way, would be all the rage with children is a small scale wind generator with the technological capacity to revolve even at slight winds and to save energy not through high power, but thanks to extended operation.

Usually, the wings of conventional wind generators are, much like those of a hawk or black kite, of a streamlined shape, and the capacity to generate electricity are proportionate to the radius of the rotor to the power of two and the speed of wind to the power of three, and thus the wings have continued to become ever larger, with some of the largest generators now on the market boasting a wingspan (diameter) of more than 150 m. The wings of such generators revolve at a speed that at the tip of the wing easily exceeds 200 km/h.

Even small wind generators on the market today require some 2–3 m/s of wind to operate. With such requirements, the time of operation per day is low, and there is no chance such generators would become the target of children's excitement. In order to make wind generators that revolve even a breeze of below 1 m/s, it is necessary to design wings of a shape that creates buoyancy (ability to revolve) even in a slight wind.

When we knock nature's door, the dragonfly appears on the other side. The dragonfly can fly through the air at lower speed than any other insect, which means it is able to turn even a slight wind into uplift. If we can learn from this mechanism, it should become possible to design generator wings that will revolve even in a slight breeze. The weight of a dragonfly is only some 1–2 g, and the air to a dragonfly is much like a viscous starch syrup. To be able to fly slowly through the air, the dragonfly has some amazing mechanisms up its sleeve. Why is a cross-section of a dragonfly's wing jagged? Why does its wing not have a smooth, streamlined surface such as that found in the hawk or black kite? Recently, light has finally been thrown on this mechanism by the Obata Laboratory at Nippon Bunri University. It was discovered that whereas the black kite or hawk have wings with a streamlined cross-section, efficient when flying at high speed, the dragonfly has wings of a jagged or rugged shape, which are more efficient when flying at low speed.

It is well known that even the most advanced jet aircraft would be completely unable to fly if it were reduced to the size of a dragonfly without a change in shape. The reason is that the air would "stick" to the wings (the so-called Reynolds number (Re) would be low, and the viscosity of the air would become dominant. Generally speaking, air craft fly at above $Re = 1 \times 10^6$, and the region around 1×10^5 is called the wall of viscosity. Below this speed, the streamlined shape of air craft wings no longer functions. The dragonfly can fly even at $Re = 1 \times 10^3$ where the viscosity of the air is dominant) and no longer be able to flow smoothly along the wings. The dragonfly, however, can fly smoothly even though it is small and is very skilled at doing so. Why? Because the wings of a dragonfly are not of the streamlined shape found in birds, but takes the shape of thin plates folded in rugged form. The dragonfly utilizes this uneven, jagged surface to create small whirlpools of air that travel from the tip of the wing inwards. This continuous stream of whirlpools, it has been discovered, works like ball bearings which form an airfoil profile and thus speedily transports the outer layer of air backwards more or less like on a conveyor belt. This is an ingenious mechanism. Conversely, with camber wings—one type of streamlined wing—the air already separates from the surface around the center of the wing (Fig. 12.4).

From understanding of this mechanism, a wing was developed with a cross-section as shown in Fig. 12.5.

The specific characteristics of the developed wings are described in Fig. 12.6. With an angle of attack between 0–12, air flow along the corrugate wing both at

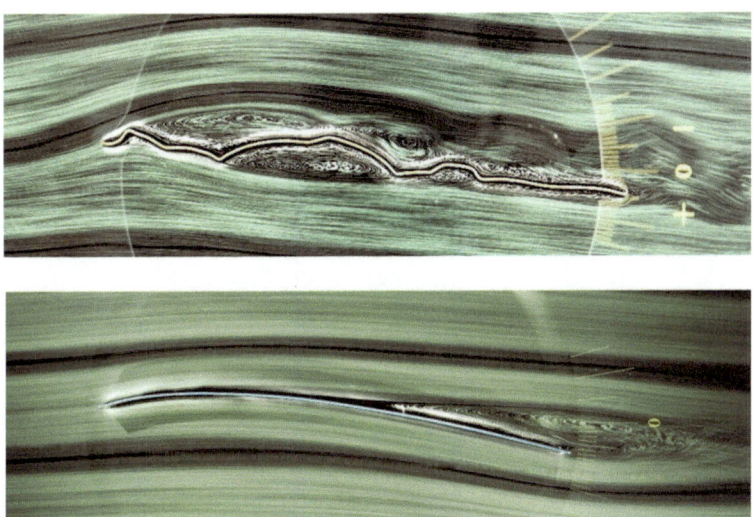

Fig. 12.4 Simulation results of airflow around a wing (Re=7,000 α (angle of elevation) =5°. Upper picture: Model of cross-section of a dragonfly wing. Lower picture: A camber wing

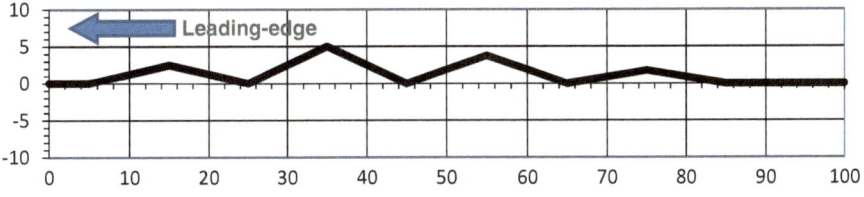

Fig. 12.5 Cross-section of a wing developed for small scale wind generators. (Corrugated wing No. 5)

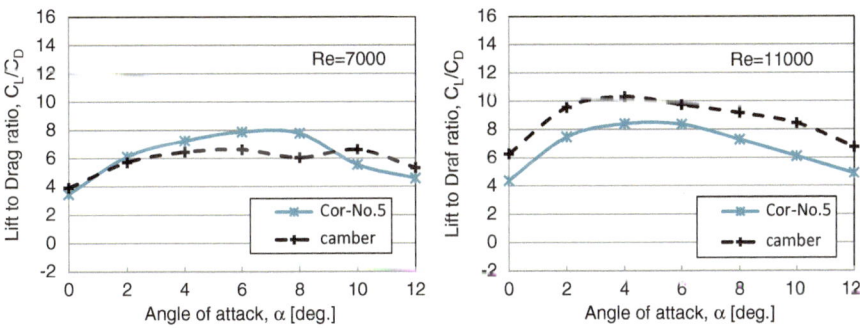

Fig. 12.6 The specific characteristics of the developed corrugate and camber wings. *Left*: Re=7,000, *Right*: Re=11,000

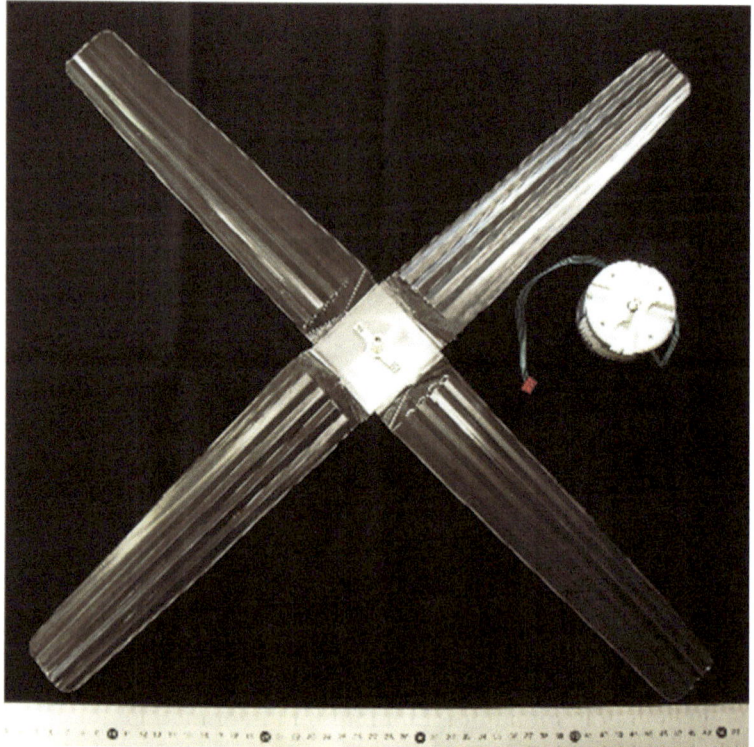

Fig. 12.7 Wing with a diameter of 40 cm

Re = 7,000 and Re = 11,000 changes smoothly in the range of a lift to drag ratio of 4–8, but, in comparison, the camber wing, said to be relatively stable at low Reynolds numbers, at Re = 7,000 displays inflection points in the range of a lift to drag ratio of 4–7, and at Re = 11,000 in the range of 6–10, thus indicating that the camber wing is more sensitive to changes in Re and lacks stability.

Based on this, a small scale wind generator with corrugate wings which are stable even at low Reynolds numbers was created. Figure 12.7 shows a wing with a diameter of 40 cm. The micro wind generator developed with this wing (Fig. 12.8) starts revolving at a wind speed of 20 cm/s and at 1 m/s is fully capable of generating electricity. Figure 12.9, shows some performance curves of this generator which indicate that at a wind speed of less than 1 m/s, this generator achieves a power coefficient of 18 % (meaning that it converts 18 % of wind energy into electricity), something conventional generators are completely unable to do at such velocities.

Another characteristic of the dragonfly's wing is that although it performs well at low speeds, the performance drops at high speeds. This is the exact opposite of what we see with conventional air craft wings. As a result of this mechanism,

Fig. 12.8 Micro wind generator

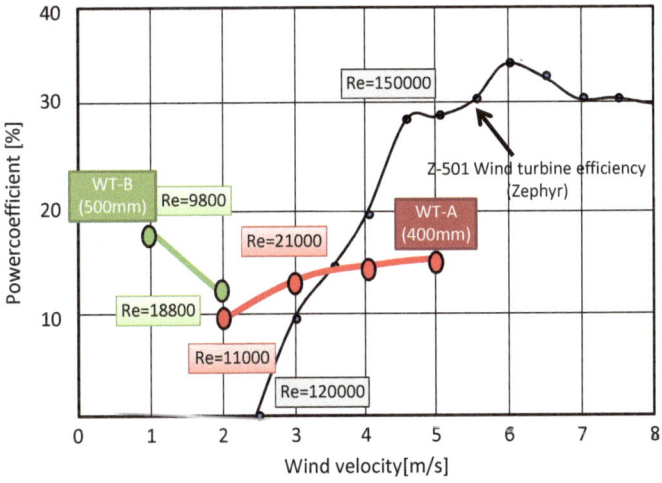

Fig. 12.9 Performance curves of micro wind generators with a diameter of 40 cm, 50 cm. A generator with 50 cm diameter wings displays a power coefficient of 18 % at wind velocity of 1 m/s. The black line indicate a small scale wind generator already on the market which has wings of streamlined (conventional) shape

we can envision a new wind generator which generates power even at low wind speeds, but which, when the winds increases, slowly lowers its performance letting the wings flutter as it moves toward a stable, low speed. Thus, it does not need the complex control systems found in conventional generators to lower speed and

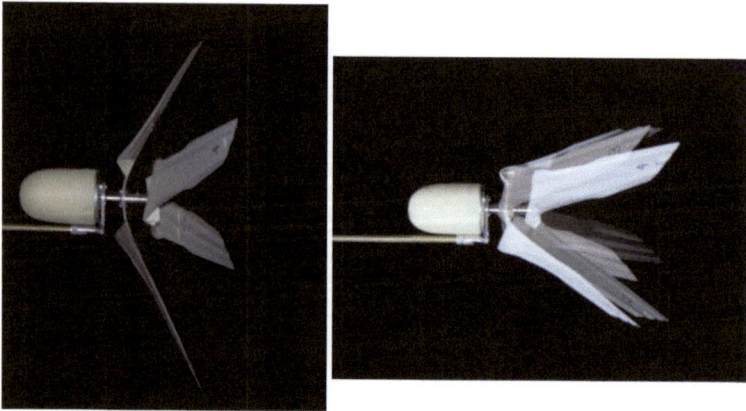

Fig. 12.10 Micro wind generator which lowers its performance when wind velocity increases. *Left*: Wind velocity 10 m/s. *Right*: Wind velocity 30 m/s (stroboscopic photography). When the wind becomes strong, the wings start fluttering and the number of revolutions per minute automatically falls

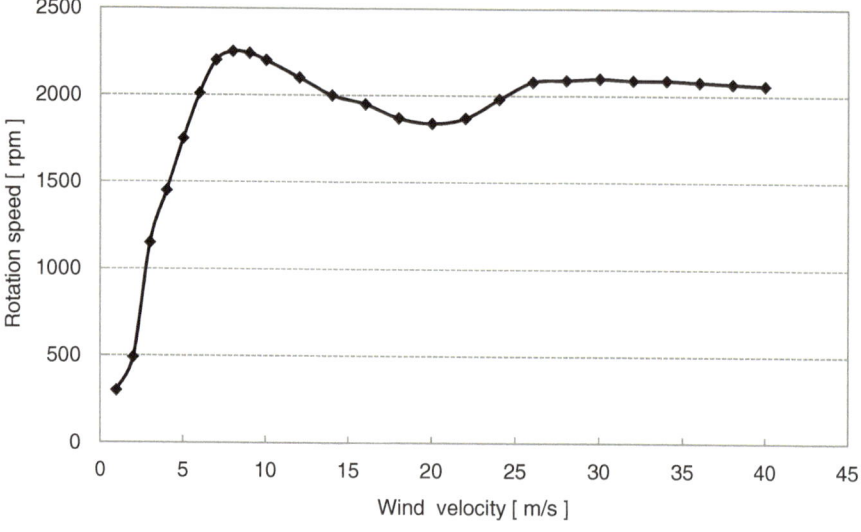

Fig. 12.11 Micro wind generator—the relationship between wind velocity and revolutions per minute

stop generator revolutions in strong winds and may withstand even typhoons (Figs. 12.10, 12.11).

In modern society, higher speed and greater power have been regarded as an unconditional good, but in nature other kinds of technologies exist, whose performance is low at high speeds but extremely high a low speeds.

12.4   A House Farm (Kitchen Garden) Learning from the Diversity of Microorganisms

The food self-sufficiency rate of Japan is 39 % (on a caloric basis, 2010. In monetary terms 69 %), and most of the imported food comes from China and some other specific countries. One can only hope it will not happen, but what if, in 2030, food imports to Japan were to come to a halt? Would the country be able to manage? The suspension of food imports is actually not as unrealistic as it may sound. It became clear in 2006 that the world had already passed peak oil, and the price of oil since then has continued to increase steadily. Meanwhile, Japan's food mileage (the transport distance of food) at 7,093 ton-km/person (2001) is by far the highest in the world, and it will become increasingly difficult to continue importing food from distant countries using huge amounts of energy. What is more, since most of the countries from which Japan imports are developing countries where rapidly growing populations make it an urgent task to secure food, there is little room left for exporting. Also domestically, the Tohoku Region, which was one of breadbaskets of the country, suffered serious damage in the Great East Japan Earthquake, leading to a situation in which rice for two million people annually can longer be produced, and at the same time, 95 % of the fishing fleet in a region providing 13 % of Japan's fish was devastated by the tsunami. The former Food Agency once estimated that even if food imports were suspended, it would be possible to supply all Japanese with 2,020 kcal/day, but this would lead to a dinner menu consisting of, for example, one bowl of white rice (75 g), one fried sweet potato (100 g), and one slice of fried fish (84 g). For breakfast or lunch, different kinds of potatoes would be the main serving; *miso* soup would only be served every other day; milk could only be had every sixth day (one cup), one egg a week; and when it comes to meat, only one serving every nine days.

Apparently, it would be possible to survive, but under such constraints, how would people be able to enjoy wholesome, fulfilling lives? New approaches would be needed to avoid getting tired of the same food being put on the table everyday (in the Edo Period, a book called "Hundred Delicacies of Sweet Potatoes" contained 123 ways of cooking sweet potato (*satsumaimo*), and new ways of farming at home would become necessary—as would, for example, common kitchens were people could enjoy farming and eating; new ways of living in which the origins of food culture are explored; and ways of storing food that do not use energy.

If the estimations of the Food Agency are correct, people would, to take one example, no longer need refrigerators at home. The reason is that fresh or raw food such as meat and fish would no longer be eaten regularly; but on the other hand, new skills for storing potatoes would become indispensable. At 13 °C, with air moisture at 90 %, it is possible to store sweet potatoes for as long as 1 year, and the question is how to realize this without using electricity. It is our plan to start knocking nature's door to look for answers.

Let us a look at one example of a lifestyle which would help enable such new ways of dealing with food.

12.4.1   *<Lifestyle Example III. A New Kitchen Garden: Growing Lettuce from the Walls and Cabbage in Drawers>*

This morning, we woke up again with sunlight pouring into the room. We live on the 15th floor of an apartment complex, but our house is full of greenery. In this season, tomatoes, cucumbers, aubergines and *goya* bitter melons are ripe and ready. For breakfast, we picked tomatoes and cucumbers from the wall and made a salad. Fresh from the plant, they are very tasty. We also picked some fresh basil from the wall in our storeroom to season the salad with. The smell is wonderful. There are so many tomatoes right now that we cannot eat them all, so yesterday we gave some of them to our neighbors. In return, they gave us some very savoury cucumbers. How can they possibly make such delicious cucumbers—our neighbors really are experts at growing cucumbers. The people living across from us even grew watermelons this year. All year round, many different vegetables grow out of the wall. One day, we must ask our neighbors to teach us how to do that.

Our neighbors are all masters at growing vegetables or flowers. It is great fun to bring some of the vegetables you have grown to a gathering and chat with the neighbors, and we have come to enjoy watching the vegetables grow inside the house day by day. Did you know that when flowers open up they give off a small sound? It is almost as if they were whispering, "I am going to start bearing big fruits".

We named the technology which enables such a lifestyle "a home farm". It is way of farming that we are hoping to popularize.

By letting plenty of sunlight enter the house, the wall can become a vegetable farm. In order to make this possible, it must be vegetables that can grown well with few nutrients and little water, and considering the fact that this takes place indoors, disinfection with chemicals must also be unnecessary. So, how to realize this? Here, again let us knock nature's door. In nature, most plants grow vigorously without having to add any nutrients. This is thanks to the great diversity of microorganisms.

It is microbial colonies existing in vast numbers in the soil which are responsible for continuously supplying nutrients to plants as well as for preventing the explosive growth of pathogenic microbes, amongst other tasks. In healthy soil, where the diversity of microorganisms is high, all forms of organic material are decomposed by microorganisms in a highly competitive environment. It is therefore difficult for pathogenic bacteria to thrive snugly. In the same way as diversity is essential in most aspects of life, greater diversity also helps make soil healthier. Therefore, if we can create a light soil (soil has a specific gravity of higher than 2. In order to grow vegetables from walls, artificial soil with a specific gravity of much less than 1 is needed) by extracting microbes from soil where they are abundant and add them to artificial soil in such a way that the diversity is preserved, the home farm becomes realistic. Furthermore, if such a system is completed, it is no longer impossible to imagine a mobile farm which could be taken into space, since plants grow with just a little water and some sunlight and heat.

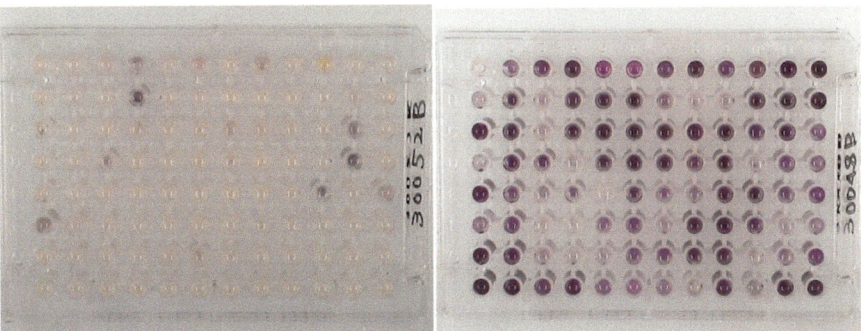

Fig. 12.12 Test plates for analyzing microbial diversity and vitality value. (DGC Technology Inc.). *Left*: Microbial diversity and vitality. Value: 246,521 (Japanese mustard spinach grown with chemical fertilizer). *Right*: Microbial diversity and vitality. Value: 1,462,275 (Japanese mustard spinach, grown with compost). The test plate with more colored samples is the one in which a larger amount of organic material has decomposed. Deeper color indicates higher speed of decomposition (result after 48 h using a special purpose robot)

There is, however, a fundamental problem when trying to create such soil. In one gram of soil, there may be hundreds of millions, or even trillions, of microorganisms, but counting the number of microorganisms or identifying what kinds are present in a particular sample of soil is close to impossible. The very act of evaluating the diversity of microorganisms is, thus, highly problematic.

Dr. Yokoyama and others from the National Agriculture and Food Research Center developed an index for soil microbial biodiversity (using the colony method), and then recently, in collaboration with Dr. Sakuramoto of DGC Technology Inc. improved this to come up with a value for microbial diversity and vitality of soil. This is an epoch-making discovery of a scientific way to measure the biodiversity of soil, hitherto difficult to evaluate. More specifically, a suspension liquid of a soil sample is added to test plates containing 95 different kinds of organic material (food for the microbes), and the way in which the organic material decomposes is measured. With a special purpose robot, the speed at which different types of organic material decompose is examined. If many different types of organic material can be decomposed, this indicates that many types of microbes are present. Also, the higher the speed of decomposing, the more active the microbes. In this way, it was made possible to come up with a value for both the diversity and vitality of microbes in soil within just 48 h (including preparatory activities, the whole exercise takes about a week) (Figs. 12.12, 12.13). The research so far has revealed that in soil where the diversity and vitality level of microbes are high, replant failures are less likely to occur even if the soil is not disinfected; residue of nitrate-nitrogen is less likely to remain in the produce even with organic farming; and the soil is less likely to attract diseases. Furthermore, this value also correlates with the hardness of the soil; that is, soil in which the diversity and vitality value is high is softer and less likely to impede the growth of roots (Fig. 12.14).

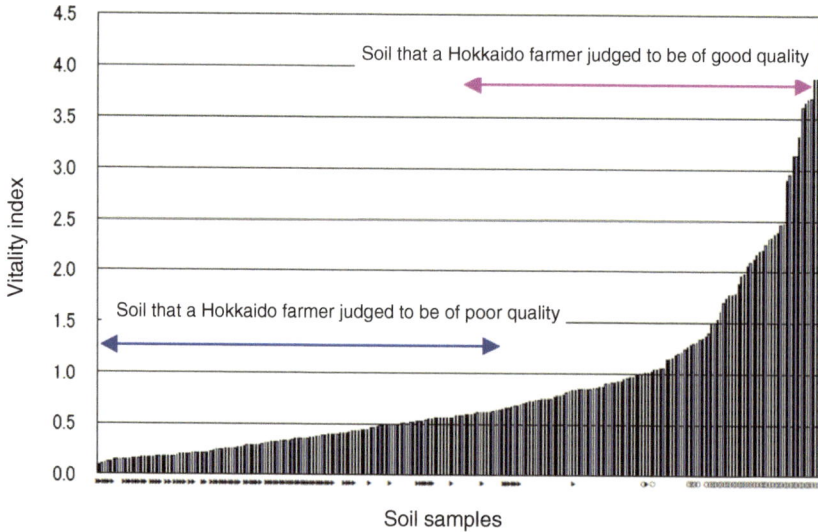

Fig. 12.13 Evaluation of soil—which used to rely on the gut feeling and experience of the farmer—can now be quantitatively conducted in short time (48 h). (*Source*: DGC Technology Inc.)

Fig. 12.14 Plants starting to grow in light-weight artificial soil abundant in microbes

It has already become clear that with ultra-light artificial soil with a specific gravity of about 0.1, it is possible to maintain microbial diversity. The home farm is thus no longer merely a dream, but rather a realistic possibility.

Growing cabbage in the living room, lettuce in the bedroom, and basil in the storeroom....the possibilities are numerous. A rooftop vegetable garden may lead to

Fig. 12.15 Images of a home farm

new communication between grandpa and the grandchildren. It may even help reinstate grandpa as an important figure in the family (Fig. 12.15).

The examples of Nature Technology we have described in this chapter were derived from lifestyles and are technologies with no black boxes.

A final remark: It is not possible to achieve a sense of fulfillment or achievement or true satisfaction only by pursuing comfort and convenience. To move forward with the agenda described in this book, we must regain however much we can of the skills we have "outsourced" and return these to our own hands. One could say this means facing constraints. Quite often it is possible, however, to gain a sense of fulfillment and satisfaction by devising ways of turning slight hardships or inconveniences into fun. This does not mean returning to the past. With a backcasting mentality, it is possible to discover lifestyles that are wholesome and fulfilling even in the face of constraints. It is also one of the roles of technology to help incorporate such aspects of richness into lifestyles, and we believe this is the sense of ethics which technology should naturally possess. We also believe that it is the Japanese, who still have a vivid sense of nature, who may be able to create such new technology.

Bibliography

Ishida H (2001) Chikyuu kankyou to monotsukuri – katatsumuri wo kagaku suru (The global environment and manufacturing – a scientific perspective on snails), vol 29 (83). Ceramic data book, Tokyo, pp 18–23

Ishida H (2002) Soil-ceramics (Earth) self-adjustment of humidity and temperature. In: Schwartz M (ed) Encyclopedia of smart materials, Elsevier, Amsterdam, pp 1014–1029

Ishida H (2006) Neichaa tekunorojii, seishinyoku wo aoru atarashii monotsukuri wo motomete (Nature technology – towards a new way of manufacturing which encourages spiritual desires), vol 34(88). Ceramic data book, Tokyo, pp 169–172

Ishida H (2007) Shizen no sugosa wo kashikoku ikasu, mudengen eakon (Utilizing nature's amazing powers intelligently – an air conditioner without a power supply). Reform Report 215, 3, JERCO, Miyagi

Ishida H (2009) Channeling the forces of nature. Tohoku University Press, Miyagi

Ishida H (2010) Atarashii kurashi to tekunorojii wo kangaeru iinkai (Committee on new ways of living and technology) Chikyuu ga oshieru kiseki no gijutsu (Miraculous technology the Earth teaches us). Shoudensha, Tokyo

Isu N, Ishida H (2003) Jiko choushitsu tairu, interijento sozai (Tiles with self-regulation of humidity – intelligent materials). In: Gijutsu no saishin kaihatsu doukou (Most recent development trends in technology), CMC, Tokyo, pp 248–251

Isu N, Ishida H (2005a) Tsuchi wo mochiita kinou kenzai (Functional building material utilizing earth). In: Clay science, vol 44 (3). The Clay Science Society of Japan, Japan, pp 129–133

Isu N, Ishida H (2005b) Tsuchi wo mochiita kankyou zairyou (Eco-materials utilizing soil). In: Ceramics, vol 40 (4). The Ceramic Society of Japan, Japan, pp 294–296

Minister's Secretariat, Statistics and Information Department, The Ministry of Health, Labour and Welfare (2012) Comprehensive survey of living conditions. The Ministry of Health, Labour and Welfare, Japan

Ministry of Agriculture, Forestry and Fisheries (2008) Shokuryou no mirai wo egaku senryaku kaigi – Tokushuu: "Shokuryou no mirai wo tashika na mono ni" (Strategy conference envisioning the future of food supply – Special focus: "Raising the certainty of our food future"). Ministry of Agriculture, Forestry and Fisheries, Japan. http://www.maff.go.jp/j/study/syoku_mirai/pdf/message_all.pdf

Taguchi Y, Ohno K, Yokoyama K (2001) Hikeiryou tajigen shakudo kouseihou he no kitai to atarashii shiten (Expectations to and new perspectives on the non-metric multidimensional scaling method). In: Proceedings of the institute of statistical mathematics, Japanese Federation of Statistical Science Associations, vol 49, pp 133–152

Takahashi Y, Yasuda Y, Tohji K, Kaya K, Ioku K, Ishida H (2007) Atarashii kurashikata no katachi (The contours of new ways of living). Geiritsu Shuppan, Tokyo

Yokoyama K (1993) Evaluation of biodiversity of soil microbial community, Biology International special issue 29, International Union of Biological Sciences, pp 74–78

Yokoyama K (2005) Kouseisha no bunrui ga konnan na kei no fukuzatsusei wo kijutsu suru tame no atarashii gainen "Bunruigun hiizongata tayousei" (A new concept to describe the complexity of systems whose components are difficult to classify). In: Fukuzatsu Genshou Kougaku 5.5 setsu (Engineering science of complex phenomena, part 5.5), National Institute of Advanced Science and Technology (Supervision). Pleiades Publishing, Nagano, pp. 333–343